You are respectfully invited
to return this book to
Gordon C. Fowler

CASE HISTORIES IN FAILURE ANALYSIS

Other ASM books relating to failure analysis and prevention

Metals Handbook, 8th Edition:

Volume 8: Metallography, Structures and Phase Diagrams

Volume 9: Fractography and Atlas of Fractographs

Volume 10: Failure Analysis and Prevention

Volume 11: Nondestructive Inspection and Quality Control

Source Book in Failure Analysis

Metallographic Etching

Color Metallography

Prevention of Structural Failures

Nondestructive Evaluation in the Nuclear Industry

CASE HISTORIES IN FAILURE ANALYSIS

 AMERICAN SOCIETY FOR METALS
METALS PARK, OHIO, 44073

Library of Congress Cataloging in Publication Data
Library of Congress Cataloging in Publication Data
Main entry under title:
Case histories in failure analysis.
  Includes articles by various authors.
  Includes bibliographical references and index.
  1. Metals—Fracture—Case studies.   I. American
Society for Metals.   II. Title.
TA460.C33   620.1'6   79-9124
ISBN 0-87170-078-6

Printed in the United States of America

PREFACE

In 1974, it was fair to state that "the grand total of published articles pertaining to failure analysis is relatively small when compared with the coverage afforded less sensitive subjects, and thus does not reflect the industrial and engineering importance of failure analysis." In that same year, the American Society for Metals published *Source Book in Failure Analysis,* followed in 1975 by Volume 10 of Metals Handbook, *Failure Analysis and Prevention.* The tide had turned at last!

It remained for ASM to publish another and exceptionally different book on failure analysis, a book that effectively supplements its forbears while establishing a very distinctive mark of its own. *Case Histories in Failure Analysis* is a first of its kind — a comprehensive collection of case histories of *actual* failures, each prepared by a recognized authority, each delineating the significant highlights of the failure, and each *identifying the cause of failure* as derived from the available evidence. Admittedly, this book is a rarity in technical publishing, if only because of the sensitive nature of its subject matter.

The case histories selected for this book were generated over a span of ten years. The selection purposively surveys both precausative factors and the causes of failure. In terms of organization, this has been accomplished by assigning the 112 case histories to four major sections according to precausative factors of failure. These have been identified as design/processing-related, service-related, materials-related, or environment-related. On the heels of these factors follow the causes of failure. Thus, a service-related failure involving fatigue has been clearly differentiated from service-related failures due to other causes, such as hydrogen damage or corrosion. Reference to the table of contents serves to guide readers to very specific types of failure in which they may have immediate interest without extensive searching and loss of time.

Of equal importance, the book has been accorded the special treatment it deserves — including the careful reproduction of hundreds of first-class photographs of fracture surfaces, microstructural details, and special surface phenomena. The entire format of the book has, in fact, been planned to accommodate the wealth of valuable detail it encompasses.

All the case histories in this book appeared originally in the pages of the international journal *Praktische Metallographie,* published by Dr. Riederer-Verlag GmbH, of Stuttgart, Germany. The American Society for Metals extends its grateful appreciation to the editorial board of this journal, to its distinguished editor, Prof. Dr. Günter

Petzow, of Max-Planck-Institut für Metallforschung, and to its publisher for permission to organize this comprehensive collection of case histories and to present them in book form. Special thanks are due Dr. F. K. Naumann, author of *Das Buch der Schadensfälle* and recently retired from Max-Planck-Institut, who — with his colleague Ferdinand Spies — contributed a majority of the case histories. We share with them this singular opportunity to serve the common interests of the engineering community.

Paul M. Unterweiser
Staff Editor
Manager, Publications Development
American Society for Metals

William H. Cubberly
Director of Reference Publications
American Society for Metals

CONTENTS

Section I: Design/Processing-Related Failures

Fatigue

Hydrogen Damage/Corrosion

Microstructure

Stress

Heat/Overheating

Forging

Casting

Hydrogen Damage/Corrosion

Microstructure

Stress

Heat/Overheating

Friction/Lubrication

Section III: Materials-Related Failures

Material Defects

Mixed Materials

Heat Treating

Section III: Materials-Related Failures *(continued)*

Aging

Overheating

Fracture

Trace-Element Contamination

Section IV: Environment-Related Failures

Hydrogen Damage/Corrosion

CASE HISTORIES IN FAILURE ANALYSIS

Section I:
Design/Processing-Related Failures

Fatigue Fracture and Weld

Friedrich Karl Naumann and Ferdinand Spies

Max-Planck-Institut für Eisenforschung
Düsseldorf

Thermal and transformation stresses, resulting from welding, adding up with operational stresses can result in failure. Sometimes, they alone initiate cracks in the weld or in the basic material, especially if the steel is hardenable, i. e. if due to fast cooling transformation in martensite occurs [1]. Fast cooling rates occur during welding, because the welding heat is quickly conducted into the relatively massive and cold material, especially if it is not prewarmed. These stress cracks act like sharp notches and can lower the fatigue strength. This applies to constructional welding as well as to patching and deposit welding. In the following are a few examples.

1. The crankshaft of a shaft-drive to produce artificial waves in a swimming pool, which was made out of a I-profile beam, was strengthened by welding strips to get a box-like profile. This failed after a comparatively short service, beginning from the flanges at a point where the strengthening ended (Fig. 1). The fracture was initiated by fatigue cracking, which started at the flanges (Fig. 2). The beam was made out of a steel, which corresponded in composition and strength the steel U St 37-2 after DIN 17 000. This was fault-free and of usual purity.

For examination of the welded joint a longitudinal section was made from the fracture origin through the flanges marked a in Fig. 2. With this many small cracks in the weld and a deeper stress-crack, which started from penetration notch and proceeded into the transformed zone of the basic material, were cut (Figs. 3 and 4). Accordingly the weld was made not only at a constructively wrong place of highest stresses, but also done faultily. This must inevitably lead to fracture.

Fig. 1. Fracture of crank-shaft, arrows = crack origination points. 0.25 x

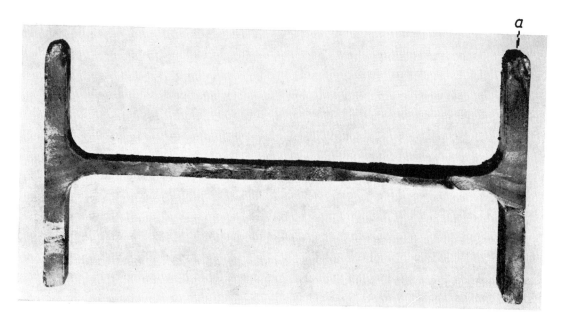

Fig. 2. Fracture of I-profiled beam. 1 x

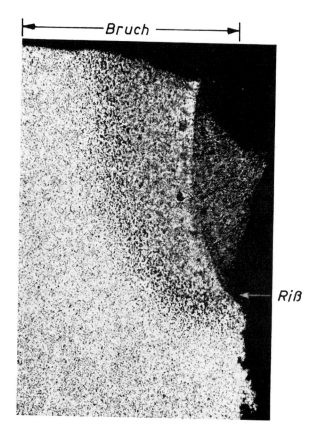

Fig. 3. Longitudinal section through a in Fig. 2.
Etching: picral. 25 x

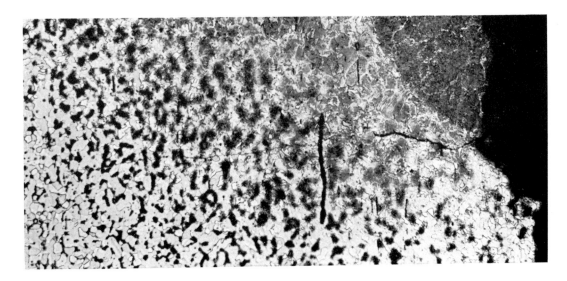

Fig. 4. Place of crack in Fig. 3. 100 x

Fig. 5. Fracture of dredger-joint bar. 1 x

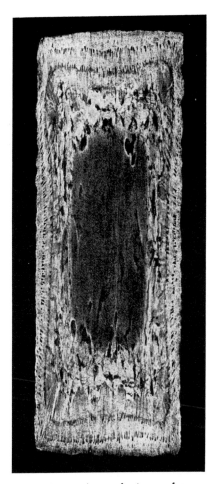

Fig. 6. Micrograph near fracture surface. Etching: copper-ammonium-chloride (after Heyn). 1 x

Fig. 7. Longitudinal section L – – L in Fig. 5. Etching: copper-ammonium-chloride (after Heyn). 1 x

Fig. 8. Place a in Fig. 7. Etching: nital. 100 x

2. The joint bar of a dredger cast out of a running non-alloyed steel with 39 kg/mm² tensile strength, which had been strengthened by welding plate strips on both sides had fractured in service. It showed a fatigue fracture, which originated at the penetration notch of the fillet weld (Fig. 5). The steel had been cast unkilled and contained considerable segregation (Fig. 6), but still had a fracture-elongation and -contraction of 68 and 75 % respectively. The longitudinal section in Fig. 7 shows that the crack runs exactly from the penetration notch to penetration notch. In Fig. 8, which is an enlarged part of Fig. 5, it can be seen that these notches are partly deep and sharp-edged. The different etchings of the welded strips in Fig. 7 show that these are not from the same or a similar plate. The whole welded construction gives an impression of being provisional rather than a planned work.

Stress- and hardening cracks were not found in this case. This failure is therefore due to a superpositioning of working and welding stresses and a stress increment due to notches.

3. An axle tube out of 40 Mn 4 after DIN 17 200 from a paper fabrication machine, which had three short longitudinal slits distributed uniformly over its surface, had cracked in two of the spaces in between (Fig. 9). The last field was broken open after dismantling of the tube. The cracks apparently started on the inner side (Fig. 10). The crack surfaces were oxidized. The last parts of the cracks on the outside are shelly and rubbed. The surface of the piece broken open later shows a normal fine grain structure with good deformation (Fig. 10 below).

Fig. 9. Side view of axle-tube. 1 x

Fig. 10. Fracture of axle-tube. 1 x

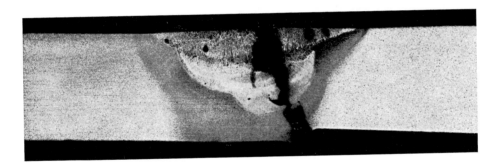

Fig. 11. Longitudinal section through fracture. Etching: picral. Outside = above, inside = below. 3 x

The shelly structure of the other two broken parts indicated the existence of a weld seam. This suspicion was strengthened because the tube had been ground in this part of the circumference (lighter strips in Fig. 9). A longitudinal section across the fracture (Fig. 11) confirmed, that the tube had been welded in both the fractured sectors from the outside, which had left an open root on the inside. The two crack edges were staggered and the resulting step had been ground down on the outside. The unwelded root corresponds to the dark strips. on the inner edge, where the fracture started.

This is a case of an unsuccessful try with unsuitable means and inadequate technique to patch up old cracks. Probably these were fatigue cracks, which originated from the sharp inner edges of the hole.

4. It is often tried through the welding to repair worn out bearings or fits and bring them to their original dimension. It is known [2] [3] that if the welding is not done very carefully taking into consideration all the precautinary measures, to which prima-

Fig. 12. Broken rear axle-tube, arrows a and b = crack origination points. 0.3 x

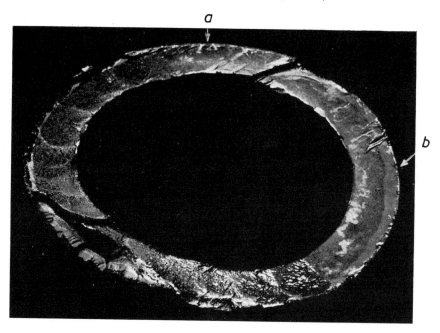

Fig. 13. Fracture, arrows a and b = crack origination points. 1 x

rily the prewarming of the work piece and slow cooling after welding belong; this try only postpones the failure.

Figure 12 shows the broken rear axle tube of a bus. The fracture is at the edge of a bearing. This is a bending fatigue fracture with several crackings (Fig. 13). One cracking (a) is at the sharp transition of cross-section and another (b) near it. A longitudinal section through the fracture showed as the cause for failure a faulty deposite welding (Fig. 14). The weld, which is lighter colored,

Fig. 14. Longitudinal section through fracture. Etching: copper-ammonium-chlorid (after Heyn). 2 x

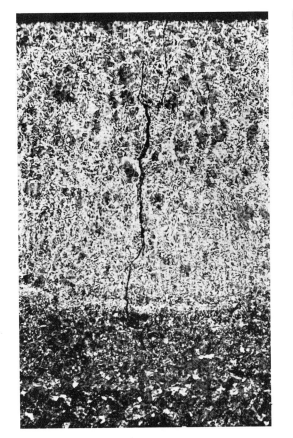

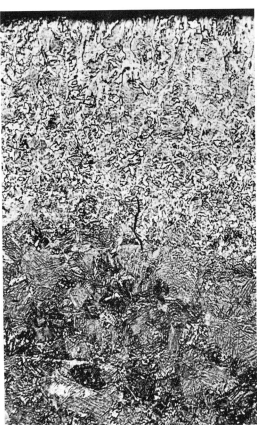

Figs. 15 and 16. Longitudinal section through the deposit weld. Edge zone with stress cracks. Etching: nital. 100 x

is full of stress cracks. The darker zone below this shows the structural change in basic material, a steel with 0,5 to 0,6 % C, which has transformed under the weld heat and has partly hardened (Figs. 15 and 16). The cracks originated in the transition between weld and basic material, where the tensile stress was highest.

[1] A. ROSE, Schweißbarkeit der hochfesten Baustähle, Einfluß der Schweißbedingungen auf das Werkstoffverhalten, Stahl u. Eisen 86 (1966) 663/672 Vgl. a./cf. F. K. NAUMANN, F. SPIES, Prakt. Metallographie 6 (1969) 747/749 und/and 7 (1970) 208/210

[2] H. SCHOTTKY, Schmelzschweißung und Dauerbruch, Kruppsche Monatsh. 7 (1926) 213/216

[3] G. KÜHNELT, Der Einfluß einer Auftragsschweißung auf die Dauerhaltbarkeit von Stahlwellen, Dissert. T. H. Berlin 1936. Masch.-Schaden 13 (1936) 57/64

Prematurely Broken Valve Spring

Friedrich Karl Naumann and Ferdinand Spies

Max-Planck-Institut für Eisenforschung

Düsseldorf

The fatigue strength of parts which are subjected to reverse bending or torsional stresses can be considerably increased by introducing compressive internal stresses into the surface zone, which is subjected to the greatest stresses. Such a state of internal stress is achieved by surface hardening, particularly by nitriding, but it can also be brought about, for example, by sand or shot blasting [1]).

The case of failure described below is intended to illustrate the fact that a faulty application of the last of the above methods can lead to just the opposite of the intended effect.

A valve spring made of 4.1 mm diameter wire, designed to withstand 10×10^6 stress cycles, fractured after only 2×10^6 cycles. It is shown in Fig. 1. The surface displayed impressions which indicated that it had been treated by shot blasting. The spring has broken in two places. Fracture 1 is a torsional fatigue fracture which has started from a lobe-like surface defect and not, as is usual, from a point on the most highly stressed inner surface (Fig. 2). Fracture 2, on the other hand, is a bending fatigue fracture with a starting point on both the inner and the outer surface of the spiral (Fig. 3). A

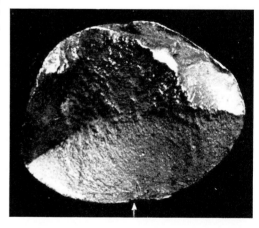

Fig. 2. Fracture 1 from Fig. 1. 10 x

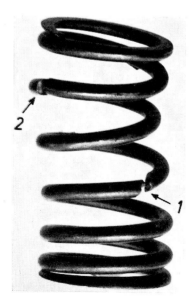

Fig. 1. Valve spring with fractures. 1 x

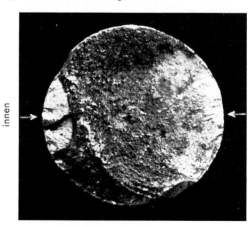

innen

Fig. 3. Fracture 2 from Fig. 1. 10 x

Fig. 4. Cracks on the inner side next to the fracture 1 of Fig. 1. 20 x

number of other incipient cracks can be seen inside and outside next to fracture 2 (Fig. 4). A longitudinal section through these cracks (Fig. 5) showed that they had started from cold worked regions, resulting from shot blasting, on the surface. In the initial stages of their propagation the cracks have followed the zone of internal shear stress, which has the form of a spherical shell, and have then penetrated into the interior across the principal stress direction. The two spring fractures have also started from such cold worked shells.

The objective of the shot blasting, to put the surface into a state of even compressive internal stress, which must first be overcome during subsequent bending and torsional loading before the boundary zone comes under tensile stress, was therefore not realized in this case. On the contrary, the shot blasting has led to a state of internal stress which has even favoured the fracture of the spring.

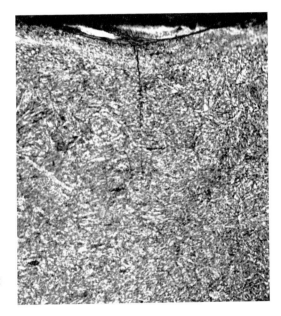

Fig. 5. Longitudinal section through cracks near fracture 2 of Fig. 1 etched in picral. 500 x

[1] A. POMP, M. HEMPEL, Arch. Eisenhüttenwes. 21 (1950) 243/262

Fractured Piston Rod of Drop Forge Hammers

Friedrich Karl Naumann and Ferdinand Spies

Max-Planck-Institut für Eisenforschung
Düsseldorf

The cause of fracture of two piston rods of hammers of a drop forge was to be determined.

1. The first rod of 180 mm diameter consisted of an unalloyed steel with 0.37 % C and 0.67 % Mn and had a strength of 56 kp/mm² at 26 % elongation.

A longitudinal section half was delivered that had already been tested for chemical composition and hardness as could be seen from a drill hole and ball impressions. Fortunately, it still showed the cause and occurrence of fracture formation. Half the fracture plane is shown in Fig. 1. The fracture origin, or rather origins of fractures were located in the interior of the cross section. Fatigue fractures propagated from several points which could be recognized as flaky cracks already in the fracture, and which later were united. No material defects could be detected in the cross section parallel to the fracture plane except for these very short cracks (Figs. 2 and 3). Figure 4 shows the cracks as small shiny cleavage fractures in a matted deformation fracture of a disk that was heat treated prior to fracture. These comparatively insignificant defects were sufficient to cause the fracture during high impact fatigue stresses in the drop forge.

2. The second piston rod of 120 mm diameter consisted of a steel with 0.25 % C and 1.00 % Mn. It allegedly had 57 kp/mm² tensile strength and 26 % elongation. Figure 5 shows that in this case the fracture originated at the surface. It was initiated by several far progressed bend fatigue fractures. The incipient fractures at the bottom of the picture are located almost in the same plane at right angles to the direction of principal stress and apparently were formed first. But

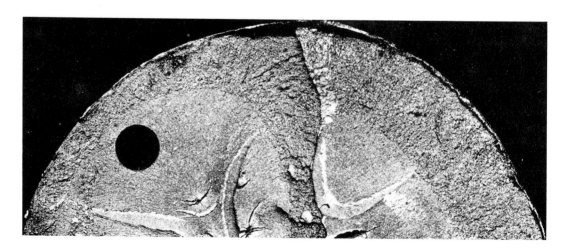

Fig. 1. Fracture of piston rod of 180 mm diameter. approx. ²/₃ x

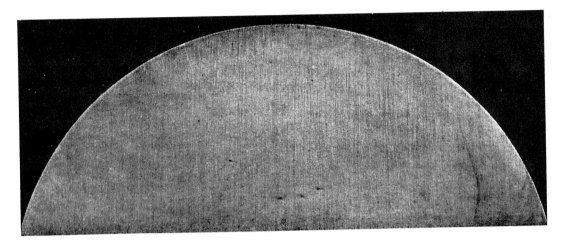

Fig. 2. Transverse section parallel to fracture, etch: Hot chloric acid 1:1. approx. $^2/_3$ x

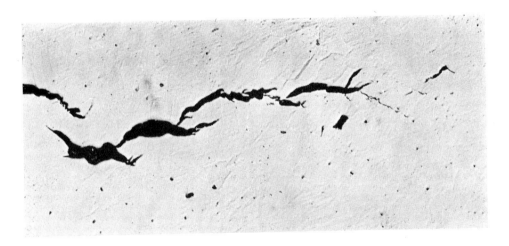

Fig. 3. Flaky cracks in unetched transverse section. 100 x

Fig. 4. Fracture of a heat treated disk. 2 x

the incipient fracture at the top of the picture is located in a cross section dislocated by 50 mm (front of picture) and runs at an inclination of approximately 45° against the longitudinal axis. The bearing surface of the piston rod was finely ground after fracture so that it could not be determined whether the occurrence of the fatigue fracture was favored by grooves or surface defects.

rich residual melt had penetrated during cooling of the block. Figure 7 shows a microsection of such a bubble. Its structure consists of manganese sulfides in phosphorous-rich ferrite. The basic structure of the piston rod was ferritic-pearlitic and hardness of 155 Brinell was accordingly low, corresponding to approximately 53 kp/mm² tensile strength.

Gas bubble segregations could be seen (Fig. 6) in a transverse section parallel to the fracture after etching with copper ammonium chloride solution and in sulfur prints according to Baumann. These are gas bubbles into which phosphorous- and sulfur-

The incipient fractures had no connection with the material defects in this shaft and therefore the fracture could not have been caused by them. Probably the low strength of the piston rod was insufficient for the high stresses.

Fig. 5. Fracture of piston rod of 120 mm diameter. approx. ²/₃ x

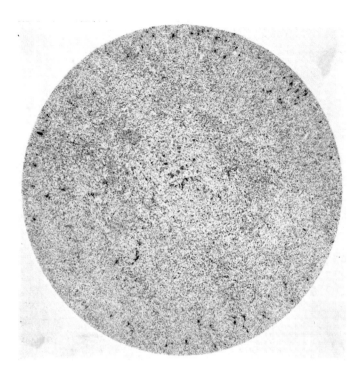

Fig. 6. Baumann print of transverse section of piston rod of 120 mm diameter. approx. $^2/_3$ x

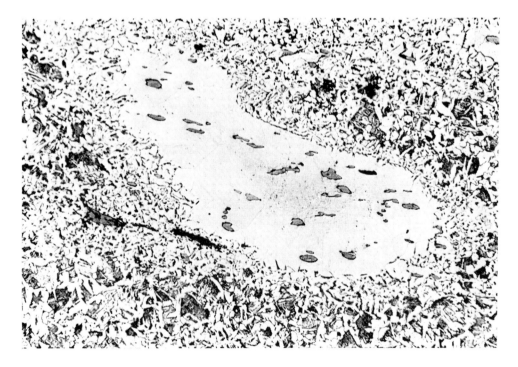

Fig. 7. Gas bubble segregation, longitudinal section, etch: Nital. 100 x

Broken Milling Machine Arbors Made of 16 Mn Cr 5 E

Gudrun Urban
Messerschmitt Bölkow Blohm GmbH
Unternehmensbereich Flugzeuge
Augsburg

The milling machine arbors were inserted with satellite spindles having a maximum speed of 1500 rpm, and broke out between the groove and the flange (Fig. 1).

1. Appearance of fracture

The appearance of the fracture surface was the same on both arbors. The pronounced scan lines characterise the fractures quite definitely as fatigue fractures. The fatigue fracture region (D) differs clearly from the ultimate ductile fracture (R) in its finer structure (Figs. 2 a and b). The course of the scan lines indicates that the origin of

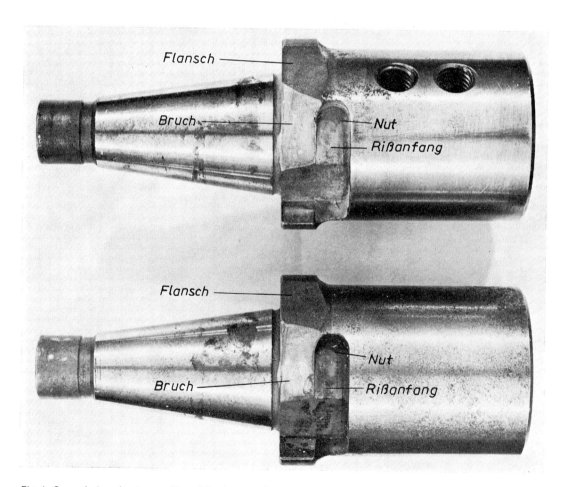

Fig. 1. General view showing position of fracture. ca. 1 x

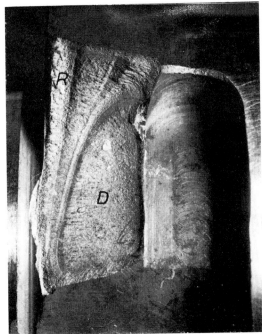

Figs. 2 a and b. Appearance of fatigue fracture (D) and ultimate ductile fracture (R). ca. 3 x

the crack is at the transition from the groove to the flange (Fig. 1).

2. Check on dimensions

The sharpness of the groove/flange transition was visible with the naked eye alone. A check on the dimensions showed that the radii were only 0.15 and 0.2 mm.

3. Hardness

Measurement of the surface hardness yielded values of HR$_c$65. The specified value was HR$_c$58—60. The hardness of the core should be HR$_c$33; HR$_c$40 was measured. Thus the surface and core hardnesses of both arbors were too high.

The case depth should have been 0.6 to 0.7 mm. The hardness curves, which were normal, showed a case depth of 0.7 mm for the limiting hardness value of HV = 550 kgf/mm² (Fig. 3). It thus lay within the required tolerance. The case depths seemed to be correct for the given cross section and range of application of the arbors.

4. Chemical composition

A check on the chemical composition confirmed that both milling machine arbors were made of the material 16 Mn Cr 5 (Table 1).

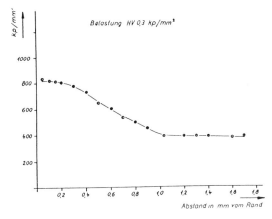

Fig. 3 Hardness curve for case

Belastung = Load
Abstand in mm vom Rand = Distance from surface in mm

Table 1. Chemical composition of the milling machine arbors in wt. %

	C %	Si %	Mn %	Cr %
Sollwert Theoretical value	0,14 0,19	0,15 0,35	1,00 1,30	0,80 1,10
Meßwert Measured value	0,16	0,11	1,1	0,8

5. Microstructure

The microstructure of the core was somewhat coarsened but still corresponded to the normal condition of the material 16 Mn Cr 5 in the quenched and tempered state (Figs. 4 a and b). The microstructure of the case showed no sign of any defect which could have led to the fracture (Fig. 5). Irregularities and crack-like grooves were present in the already too small radii (Fig. 6).

6. Conclusion

The appearance of the fracture in the arbors indicated ductile fatigue fracture which had its origin in the radii between groove and flange. These radii of 0.15 and 0.2 mm were too small for the load on the milling machine. In addition there were grooves at the base of the radii which must have had an unfavourable effect on the life of the component by acting as notches with their resulting stress concentration. Considering the great hardness of the case, the small radii would have been critical even without grooves.

Measures were taken so that the critical radius of the milling machine was increased and the surface roughness more precisely measured.

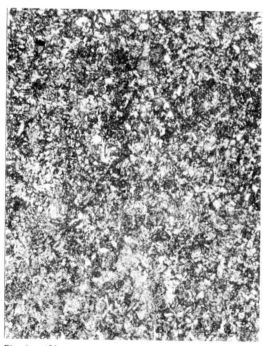

Fig. 4 a. 100 x

Fig. 4 b. 500x

Figs. 4 a and b. Microstructure of core

Fig. 5. Microstructure of case. 100x

Fig. 6. Radius between groove and flange with crack-like score marks. 200 x

Fracture of a Steering Damper

Fulmer Research Institute Ltd.
Buckinghamshire, England

The piston rod of a steering damper on a single decker bus fractured, after 100,000 miles service, in the fully extended left full-lock position. The steering damper, which is similar in shape and operation to a telescopic shock absorber, is secured by ball joints locked with slotted nuts. The steel piston rod fractured at the axle end leaving approximately 5 mm of rod welded to a securing ferrule. An exploded view of the failed damper is shown in Fig. 1.

Inspection of a new damper on the vehicle from which the fractured one had been removed showed that the damper did not vibrate nor move under low and high engine revving in neutral gear. It was found that in the left full lock position the damper cannot over-extend to give an impact blow since a wheel stop is encountered by the steering arm 10 mm before the damper's piston becomes fully extended.

The fracture surface of the ferrule end of the piston (see Fig. 2) was mainly flat apart from a raised lip, approximately 1 mm high, the top of which coincided with a right angled shoulder machined around the rod to locate the end dust cap. Beach markings, typical of fatigue, were present on one half of the flat fracture area indicating that a crack started at the surface position arrowed "B" in Fig. 2.

A ring of cratered metal was present around the base of the piston shown in the foreground of the figure. This extruded weld metal formed during resistance welding between the securing ferrule and piston rod.

Longitudinal sections for metallographic examination were taken through both halves of the fracture and a similar section was taken through a used unbroken damper to include the ferrule, dust cap and piston rod.

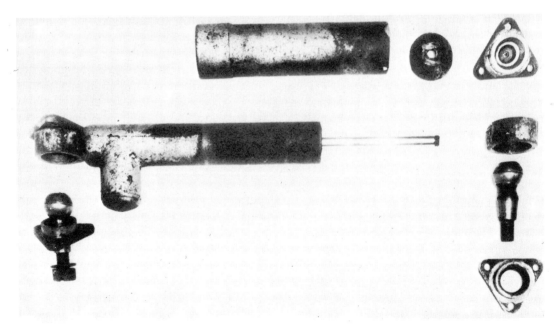

Fig. 1. Exploded view of the failed damper

Several cracks were found in the piston from the failed damper near to the weld. Figures 3 and 4 illustrate two sides of the piston half of the fracture; on one side (Fig. 3) are shown cracks running into the piston from the shoulder line and just above the unradiused shoulder towards the weld. The other side (Fig. 4) shows that the fracture path runs into the piston in the same area as the shoulder line. Figure 5 illustrates part of the ferrule half of the fracture showing the fractured lip and cracks just below it.

Fig. 2. Fracture surface of the ferrule end of the piston. A crack started at the surface position arrowed "B".

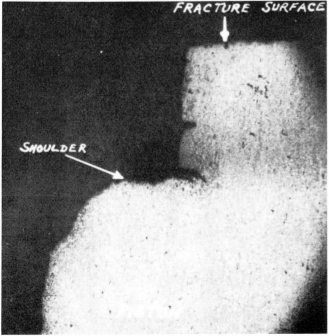

Fig. 3. Piston half of fracture showing cracks running into piston from shoulder line. 20 ×

No cracks were found in the unbroken pistons and the weld (see Fig. 6) between the ferrule, end cap and piston was satisfactory. The heat affected zone (H. A. Z.) did not extend to the unradiused shoulder of the piston. Figures 7 and 8 show that this is not the case for the broken damper and that all cracks are in the hardened heat affected zone. Figure 9 shows the ferrite/martensite microstructure in H. A. Z. near the fracture surface. This structure has a hardness of 425 HV whereas a significantly lower reading of

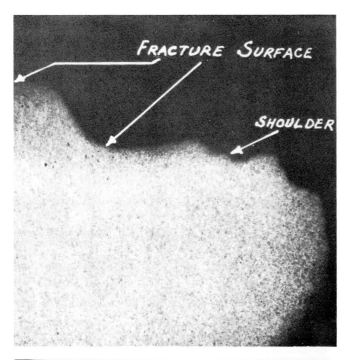

Fig. 4. Piston half of fracture — other side to Fig. 3. Fracture paths run into piston from shoulder line. 20 ×

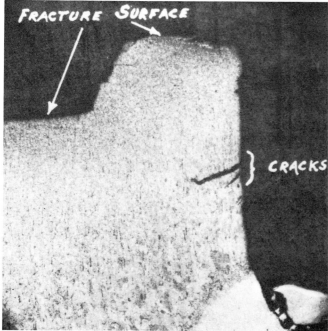

Fig. 5. Ferrule half of fracture showing fractured lip and cracks. 20 ×

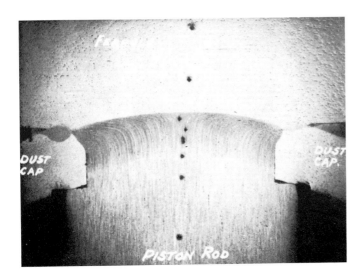

Fig. 6. Weld area of an unbroken piston — note that the H.A.Z. does not extend to shoulders and no cracks are present. 5 ×

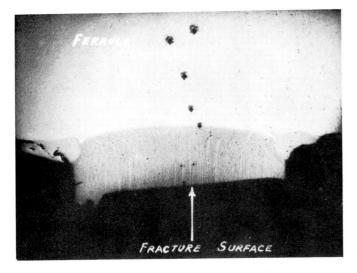

Fig. 7. Ferrule end of fractured piston showing fracture across H.A.Z. 5 ×

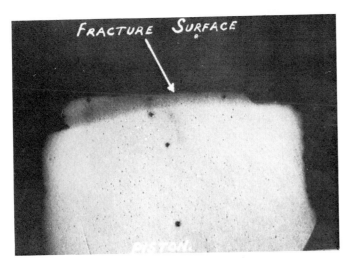

Fig. 8. As Fig. 7. but piston end showing H.A.Z. in line with shoulder. 5 ×

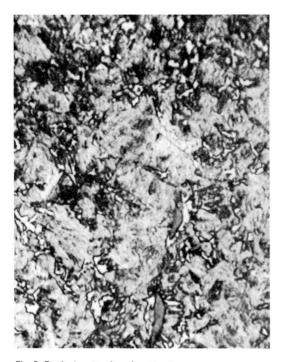

Fig. 9. Ferrite/martensite microstructure near fracture surface in H.A.Z. 1000 ×

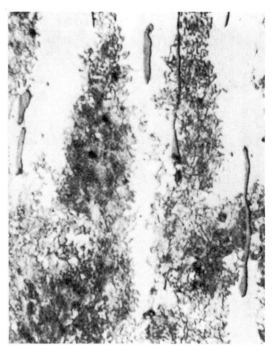

Fig. 10. Unbroken piston showing ferrite/pearlite structure at comparable position to Fig. 9. 1000 ×

250 HV was obtained from a comparable position of the unbroken piston where the structure (see Fig. 10) was ferrite and unresolvable pearlite.

The difference in hardness between the two dampers can be seen from the different size of the hardness impressions in Fig. 6, 7 and 8.

The structure of both pistons well way from the H.A.Z's were typical of normalised steels; the carbon contents were 0.37% and 0.34% for the unbroken and broken pistons respectively.

The most striking features revealed by metallographic investigation are the presence of cracks in the H.A.Z. and the depth of the H.A.Z. which reached the unradiused shoulder of the fractured piston. Both these features contributed towards failure. The H.A.Z. of the cracked piston is significantly larger than that of the uncracked piston which fell short of the shoulder line. These differences are due to a higher welding temperature and possibly a longer welding time for the broken piston.

Summary

The failure was caused by a fatigue mechanism. Small surface cracks formed during welding in the heat affected zone close to an unradiused shoulder in the piston. Under alternating stresses in normal service these cracks propagated through the piston rod made less tough by the extended weld heat affected zone.

Broken Rear Wheel Suspension

Friedrich Karl Naumann and Ferdinand Spies
Max-Planck-Institut für Eisenforschung
Düsseldorf

One of the rear wheels spun off a sports car of the racing type during a run on the express highway. An investigation revealed that the suspension bushing had broken out of the disk (Fig. 1). The circular rupture was initiated by several fatigue fractures which originated partly at the inner and partly at the outer fillet at the transition from bushing to disk (Fig. 2). Both changes in cross section were roughly finished (Fig. 3); apparently the exterior one was even of sharply-edged formation. At the torus on the outside, fresh turning grooves were present at four locations, pairs of which were situated opposite to one another (Figs. 2 and 4). A longitudinal section (a of Fig. 2) showed that at these locations a weld had been built up (Fig. 5).

The welding heat and rapid cooling caused the base material below to harden throughout the torus. The light region has a completely martensitic structure (Fig. 8), and is indeed already permeated with hardness cracks (Fig. 7). The welding material has transformed in the intermediary stage (Fig. 9). The original condition of the torus is included in section b of Fig. 2 and is shown in Fig. 6. The normal heat treated structure is reproduced in Fig. 10.

On the basis of this finding the other rear wheel suspension was disassembled and examined as well. No exterior cracks were noted but it also showed four welded regions on the torus, located at 90° intervals from one another (Figs. 11 and 12). Sections through these locations already showed stress cracks in the hardened base material of the transition region (Figs. 13 and 14).

The fracture of the one rear wheel suspension has therefore been caused by welding stresses and hardness cracks due to built-up welds on the torus on the outside of the disk. The rough finish and the sharp-edged forma-

Fig. 1. Rear wheel suspension with fracture. 0.6 x

Fig. 2. Broken-out bushing with fracture. 1 x

Fig. 3. Turning grooves and fatigue fractures in the outer fillet. 5 x

Fig. 4. Turning grooves and built-up weld in the outer torus. 5 x

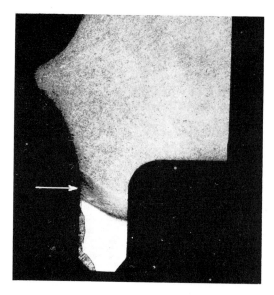

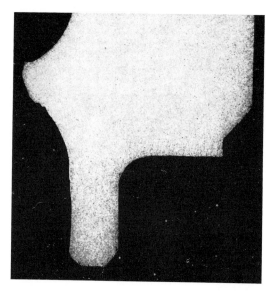

Fig. 5. Longitudinal section a of Fig. 2 with built-up weld and contiguous hardened region. Etch: picral. 3 x

Fig. 6. Longitudinal section b of Fig. 2 with unaltered structure, etch: picral. 3 x

tion of the cross section transition may have contributed to the failure. The suspension of the other rear wheel showed the same characteristics; sooner or later it would also have failed.

This event very clearly demonstrates the lack of responsibility on part of machine shops that execute or permit repair-welds on highly stressed machine parts and especially vehicle components.

Fig. 7. Longitudinal section a with built-up weld (left) and hardened region with hardness crack, etch: picral. 50 x

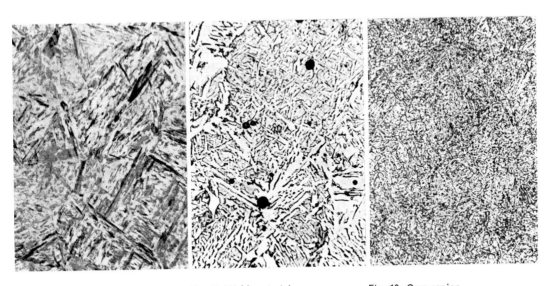

Fig. 8. Hardened region   Fig. 9. Weld material   Fig. 10. Core region

Figs. 8 to 10. Structure in longitudinal section a, etch: picral. 500 x

(Continued on the next page)

a

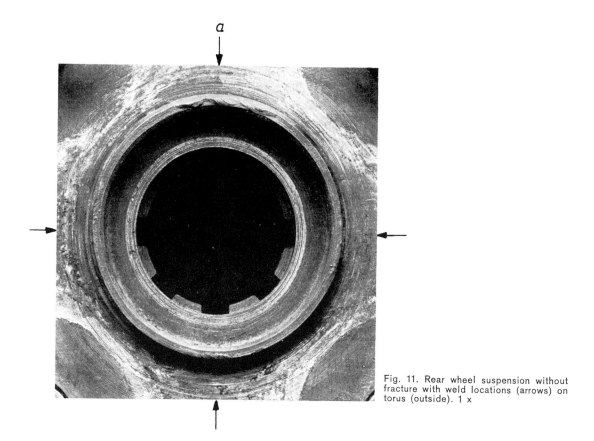

Fig. 11. Rear wheel suspension without fracture with weld locations (arrows) on torus (outside). 1 x

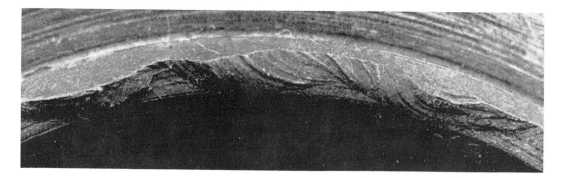

Fig. 12. Weld location on torus. 5 x

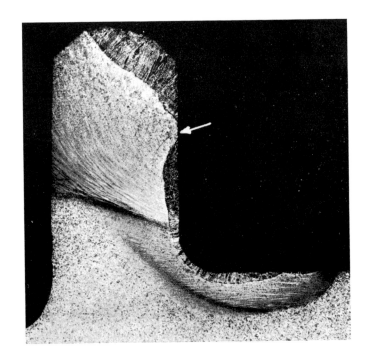

Fig. 13

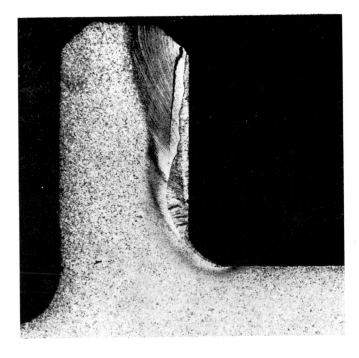

Fig. 14

Figs. 13 and 14. Sections through weld locations on torus, etch: picral. 9 x

30

Fractured Valve Spring

Friedrich Karl Naumann and Ferdinand Spies
Max-Planck-Institut für Eisenforschung
Düsseldorf

The fractured valve spring shown in Fig. 1 showed a small fatigue fracture and a large ductile fracture emanating from it in the fibrous fracture structure, that also appeared in part as a shear fracture (Fig. 2). Accordingly, stresses were evidently high. The fracture originated in the inner fiber, as usual. The fatigue fracture ran along the fiber for a short stretch (Fig. 3) and then continued as a torsion fracture located at 45° to the longitudinal axis. The fact that the fracture ran initially in the direction of the fiber pointed to the existence of a surface defect.

A transverse section adjacent to the fracture and a longitudinal section through the crack breakthrough were cut for metallographic examination. No major surface defect was found, in spite of the fact that it had been suspected according to the fracture path. But a number of minor surface defects such as rolling laps (Figs. 4 to 6) were found. Some fatigue fractures originated in part from these.

The steel itself was free of defects and was of high purity. The types and amounts of nonmetallic inclusions corresponded to the classes M < 01.1 and S ≈ 04.1 according to Steel-Iron-Test Standard 1570. The spring was heat treated (Fig. 7) and its surface was strengthened by means of shot peening (Fig. 9). The heat treated structure was normal. The austenitic grain size was smaller than group 8 according to Steel-Iron-Test Standard 1510. The surface was decarburized to a depth of approximately 0.03 mm (Fig. 8), lowering the surface hardness (HV 0.1) from 560 to 450 kp/mm².

The fracture of this valve spring is therefore primarily due to surface defects, and secondly perhaps also to weak surface decarburization. This case shows that comparatively minor effects suffice to cause fractures in highly stressed springs.

Fig. 1. Fractured spring. 1 x

Fig. 2. Fracture surface, fracture origin designated by arrow. 10 x

Fig. 3. Side view of fracture origin. 10 x

Fig. 4. Surface defect, transverse section, unetched. 500 x

Fig. 5

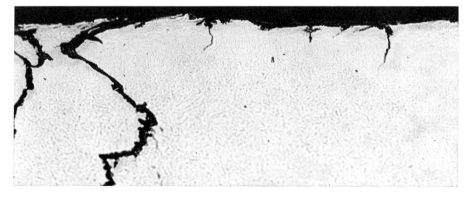

Fig. 6

Figs. 5 and 6. Surface defects with fatigue fractures, transverse section, unetched. 500 x

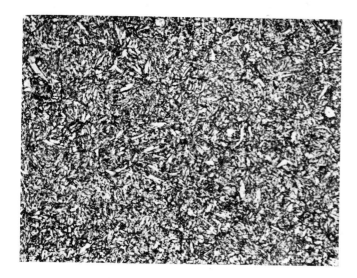

Fig. 7. Microstructure of spring. 500 x

Fig. 8. Region of surface decarburization. 200 x

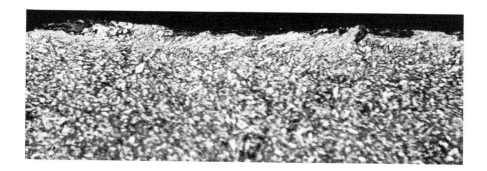

Fig. 9. Structure near surface with deformation caused by shot peening

Figs. 7 to 9. Transverse section, etch: Picric acid

Failure of Recuperator with Austenitically Welded Pipes

Friedrich Karl Naumann and Ferdinand Spies

Max-Planck-Institut für Eisenforschung

Düsseldorf

A recuperator used for preheating the combustion air for a rolling mill furnace failed after a relatively short service time because of leakage of the pipes in the colder part. Here the 6 % chrome steel pipes used for the warmer part are connected by means of welding with austenitic electrodes to the unalloyed mild steel pipe of larger diameter. Such damage did not occur in the warmer zone at similar connections.

Figure 1 shows a leaky connection. The seam is cracked on the side of the thicker mild steel pipe. It was deeply eroded over the whole circumference in a trenchlike fashion at the transition to the unalloyed pipe. Figure 2 reproduces a longitudinal section through this location. Corrosion made its appearance in close contact with the mild steel pipe and has consumed almost the entire wall thickness. In contrast,

Fig. 1. Pipe specimen with leaky weld. 1 x

Fig. 2. Longitudinal section through the weld. Etchant: Nital. 8 x

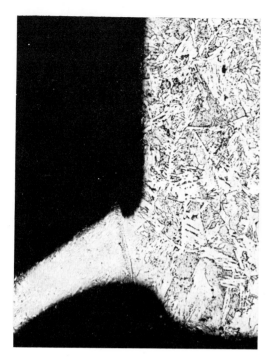

Fig. 3. Transition from mild steel pipe to welding material, longitudinal section, etchant: V2A-etching solution. 100 x

the chrome steel pipe on the other side next to the welding seam was only slightly corroded over a short distance which approximately corresponded to the heat-affected region.

Figure 3 shows the transition from the mild steel pipe to the welding material that had frozen in position. The otherwise fine-grained ferritic-pearlitic microstructure of the unalloyed pipe (Fig. 4) has become coarse-grained and acicular (Widmanstätten structure). The microstructure of the welding seam was not austenitic but had become predominantly martensitic (Fig. 5) as a result of the mixing of the weld metal with the fused pipe material. The chrome steel pipe had a structure composed of ferrite with finely dispersed carbides (Fig. 7). At the transition to the weld it had partially transformed to martensite or bainite (Fig. 6).

It is therefore a typical case of contact corrosion wherein the alloyed welding seam represented the less noble electrode.

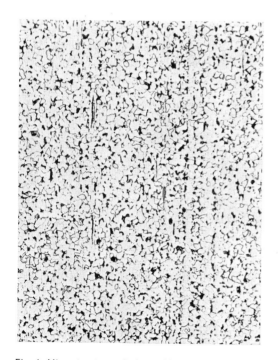

Fig. 4. Microstructure of the mild steel pipe, etchant: Picral. 100 x

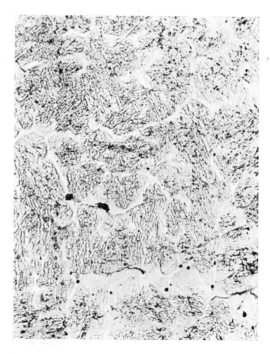

Fig. 5. Microstructure of the welding material, etchant: V2A-etching solution. 500 x

Fig. 6. Microstructure of the transition zone, etchant: V2A-etching solution. 100 x

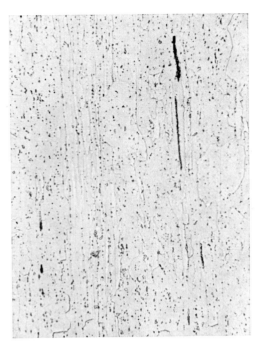

Fig. 7. Microstructure of the chrome steel pipe, etchant: V2A-etching solution. 500 x

The martensitic structure may have contributed to this in view of its internal state of stress. That the attack cannot be attributed to a sealing process is not only proven by the external appearance but also by the fact that further inside the condenser, where because of the higher temperature no condensate formed, welding seams of the same type were undamaged. An indication as to the type of the corrosive medium may have been given by the contributor of the specimens who reported that moist deposits containing 15 % sulfur were found on the damaged pipes.

Damaged Section of a Worm Drive

Egon Kauczor

Staatliches Materialprüfungsamt an der Fachhochschule
Hamburg

A general view of part of a worm drive made of corrosion resistant chromium nickel steel X2 CrNiMo 18 10 is shown in Fig. 1 as submitted for examination. The specimen comes from a worm drive used in a chemical works at ca. 80° C and 100 to 200 atm. pressure for transporting, among other things, media containing chloride. The whole surface of the specimen was covered with fine branching cracks and was flaking off in many places.

A specimen was taken for metallographic examination at the point indicated with an arrow in Fig. 1. Figure 2 shows the unetched polished specimen surface lying perpendicular to the surface. It can be seen that the surface layer has been completely broken up by numerous branching cracks. The disintegrated zone extends on average to 1 mm below the surface.

The micrograph of the etched section in Fig. 3 shows numerous deformation lines in the austenite grains. The transgranular character of the cracks can also be seen. Grain disintegration is thereby eliminated as a cause of failure, as would be expected in the case of this extra low carbon steel.

The overall pattern of the damage is typical of stress corrosion cracking. Also, and as a consequence of this, further corrosion will have occurred due to ventilation elements which were able to form beneath the flaked off material.

Stress corrosion cracking occurs when the following three conditions are simultaneously fulfilled.

1. The material must have a particular tendency to stress corrosion cracking.

Fig. 1. General view of the section of a worm drive as submitted. A specimen was taken for microscopic examination at ←. 1/3 x

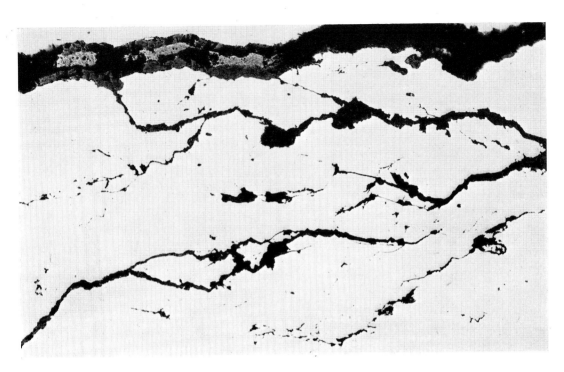

Fig. 2. Unetched microsection taken perpendicular, to the surface at the point marked with an arrow in Fig. 1. 100 x

Fig. 3. Microstructure in the cracked zone revealed by etching with V2A pickle. 200 x

2. Static tensile stresses must be present, it being immaterial whether they are applied or internal.

3. There must be an attack by some agent capable of initiating stress corrosion cracking in this particular material.

The susceptibility of austenitic chromium nickel steels to stress corrosion cracking, particularly in the presence of chlorine compounds, is well known. At high temperatures, as in the present case, it can be initiated by the smallest quantities of chloride. The deformation lines in the microstructure indicate that internal stresses are present in a thin surface layer resulting from cold working during machining. The drive was used in the works to transport media containing chloride.

Stress corrosion cracking is no longer to be reckoned with if one of the three factors can be eliminated. Owing to their microstructure, the austenitic rust- and acid-resistant steels tend particularly strongly to smearing and hence to work hardening during machining. In order to reduce work hardening, therefore, these types of steel should be machined only with sharp, short tools mounted in rigidly built machines.

Residual stresses can however be eliminated with certainty only by annealing after machining, if possible in a protective atmosphere. Since a molybdenum alloy austenitic steel is involved here, the annealing temperature should not lie below 1000° C, preferably between 1050 and 1100° C, followed by rapid cooling in air, in order to avoid embrittlement by sigma phase precipitates.

Failure of a Weld Seam in a Heat Exchanger of an Ammonia Synthesis Plant

Friedrich Karl Naumann and Ferdinand Spies

Max-Planck-Institut für Eisenforschung

Düsseldorf

A heat exchanger was removed from an ammonia synthesis plant after five years' service because loss of gas indicated a leak. The container consisted of a number of hollow cylindrical forgings which were welded together. Alloy steel of designation 12 CrMo 19 5 (Material No. 1.7362) [1], which is highly stable in hydrogen, was used. The temperature reportedly did not exceed 400° C at a hydrogen partial pressure of approx. 600 atmospheres. Under these conditions the steel that was used can be considered absolutely stable. It was therefore suspected that a weld seam was leaky or had become so. A fine fissure was indeed located by X-ray examination in one of the welded seams. A specimen was removed from this location for investigation (Fig. 1).

An etched cross section through the round seam is shown in Fig. 2. The fissure originated on the outside, at the transition from the unetched austenitic welding material, which is highly resistant to attack by hydrogen, to the basic material of the container. For some distance, the fissure followed the transition

zone, after which it deviated in a direction perpendicular to the inner wall of the tube, without however reaching it at the examined location.

Figures 3 and 4 reproduce a section from the flawed region at a higher magnification. This reveals an approx. 0.5 mm wide zone adjacent to the weld seam which is permeated with intergranular cracks as a result of hydrogen attack. The fissure originated in this zone but later left it in order to follow the direction of maximum stress. Figures 5 and 6 show the structure adjacent to the weld seam at higher magnification. The zone incorporating the grain boundary cracks is completely martensitic. The weld seam was therefore not tempered or annealed. This is also evident from the hardness values shown

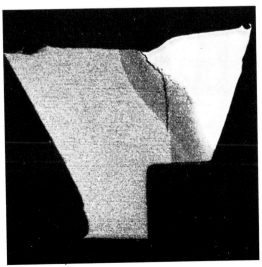

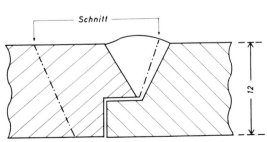

Fig. 1. Position of the specimens in the wall of the container

Fig. 2. Longitudinal cross section, see Fig. 1, etch: picral. 4 x

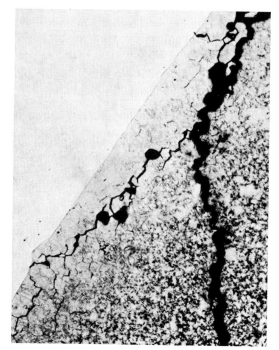

Fig. 3. Unetched
Figs. 3 and 4. Longitudinal cross section, see Fig. 2. 50 x

Fig. 4. Etch: picral

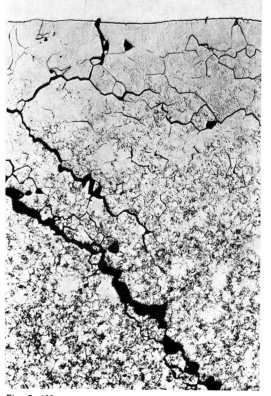

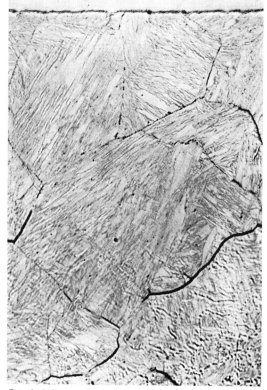

Fig. 5. 100 x

Fig. 6. 500 x

Figs. 5 and 6. Martensite zone adjacent to the austenitic weld seam, etch: picral

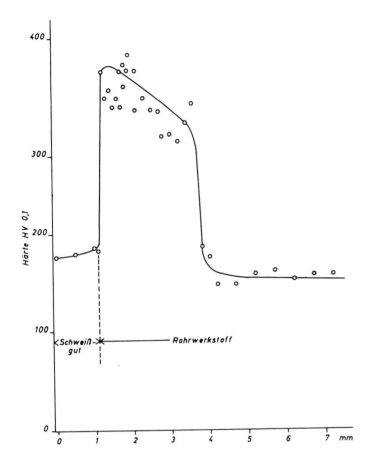

Fig. 7. Hardness values near the weld seam

in Fig. 7. Whereas the weld has an approximate hardness of 180 (HV 0.1) and the base material one of 150 kg/mm², the hardness in the critical zone next to the seam increases to 380 kg/mm².

As mentioned previously [2]), steel can be made resistant to attack by hydrogen under pressure by the addition of elements which bind all of the carbon in the form of stable carbides of low solubility. This should be the case in the presently used alloy containing 5 % Cr and 0.5 % Mo. The equilibrium phases and concentrations will however only appear if the steel is cooled slowly from the austenitic region. This is usually not the case

after welding. In this case the unstable iron carbide may segregate as a transition phase in a zone immediately adjacent to the seam, since this zone is heated to the highest temperature and cooled at the fastest rate. This unstable phase may be transformed into the equilibrium carbide by means of tempering or annealing.

The damage therefore arose because of the omission to anneal the container, or at least the welded regions, after the welding process.

[1]) Werkstoff-Handbuch Stahl und Eisen, 4. Aufl., Blatt 085 u. Stahl-Eisen-Werkstoffblatt 590
[2]) F. K. NAUMANN, F. SPIES, Prakt. Metallogr. 8 (1971) 375/384

Steel Sliver in a Continuously Cast Aluminum Press Stud

Friedrich Karl Naumann and Ferdinand Spies

Max-Planck-Institut für Eisenforschung
Düsseldorf

In a continuously cast aluminum press stud two small foreign metal slivers were found that had caused difficulties with the cable sheathing press. An investigation was conducted to find out whether the slivers were present in the melt or whether they penetrated the metal from the outside during working.

For the metallographic examination the slivers were polished and etched first with Nital and subsequently with 1:10 sodium hydroxide. As can be seen from Figs. 1 a and b, the slivers consisted of a core that

was etched by Nital, and a metallic case that was not etched by Nital, but was dissolved by sodium hydroxide. Under the microscope the core showed the structure of a heat treated alloy steel (Figs. 2 a and b). A strongly fissured and therefore brittle alloy zone was in contact with the steel. The existence of a superlattice phase Fe_3Al was probable. At the outer edge metal was adhering at some places that was roughened during polishing and therefore was soft (Fig. 2 c). During treatment with sodium hydroxide or alkaline sodium picrate solution the alloy zone was etched (Figs. 3 a and b) and the metal completely dissolved.

Fig. 1 a. Etch: Nital

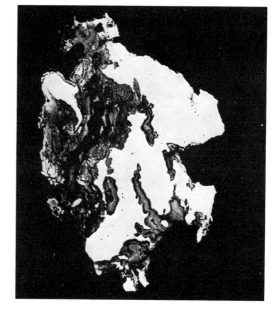

Fig. 1 b. Etch: Sodium hydroxide 1:10

Figs. 1 a and b. Section through a sliver. 10 x

Figs. 2 a to c. Structure of a sliver, etch: Nital

Fig. 2 a. View. 100 x
Top (rough and spongy): metal
center (fissured): alloy
bottom (etched): steel

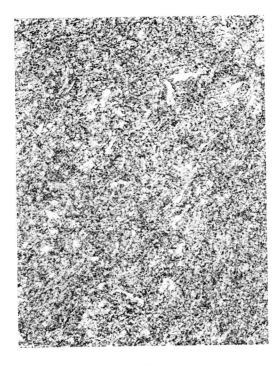

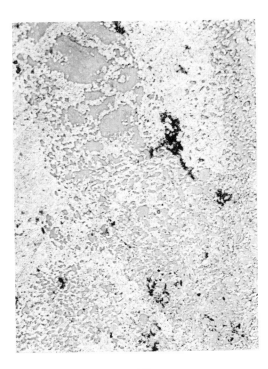

Fig. 2 b. Steel structure. 500 x

Fig. 2 c. Metal structure. 500 x

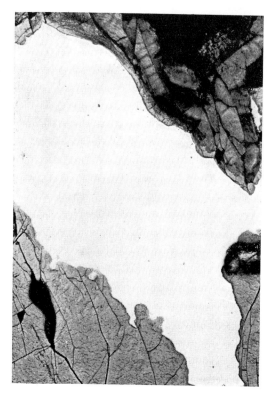

Fig. 3 a. View. 100 x
unetched: steel
etched: alloy

Fig. 3 b. Alloy. 500 x

Figs. 3 a and b. Structure of a sliver, etch: alkaline sodium picramate

A small load hardness test (HV 0.1) gave the following hardness values for the individual microstructural zones. They are not absolute values because of the load dependence, but may be compared with each other:

Heat Treated Alloy Steel	276	285	274	247
Alloy Zone		782	782	772	742
Metal		27	147	27

Spectroscopic examination revealed that the slivers consisted of a chromium-molybdenum-vanadium steel with minor (unintentional) additions of copper, nickel, and cobalt. Microanalysis after dissolution of the aluminum-containing zones gave the following values:

Cr %	Mo %	V %	Al *) %
5,56	1,19	0,28	0,37

*) from remaining adhesions

A steel of similar composition, the material X 38 Cr Mo V 5 1 (W-No. 2343) is used for hot working tools.

This result is strong proof that the sliver originated from a damaged press tool.

Metallic Inclusions in Steel

Friedrich Karl Naumann and Ferdinand Spies
Max-Planck-Institut für Eisenforschung
Düsseldorf

Occasionally metallic inclusions are found in ingots, semi-finished or finished products which top crusts resemble[1] [2] [3]. While these can be differentiated from the steel in the primary structure but not in composition or the secondary structure, the former may have different basic composition and therefore also a different structure than the surrounding material[4]. How such foreign matter has entered into the cast structure is usually impossible to establish later. For the nowadays rarely used stirring of castings or pour-

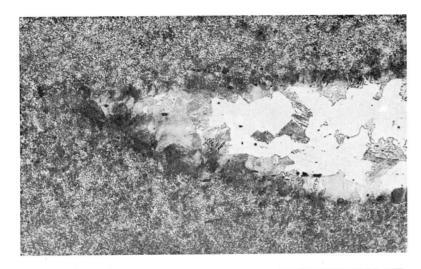

Fig. 1. 100 ×

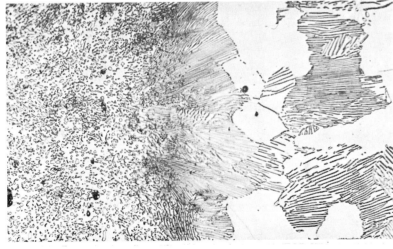

Fig. 2. 500 ×

Figs. 1 and 2. Foreign inclusion in an annealed piece of chromium steel with approx. 1 % C and 1.5 % Cr. Longitudinal section. Etch Picral

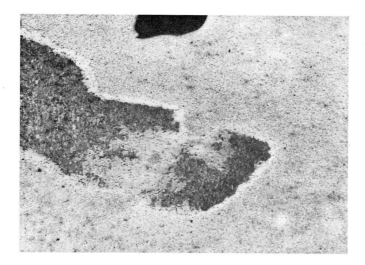

Fig. 3. 25 ×

Fig. 4. 500 ×

Figs. 3 and 4. Foreign inclusion in a hardened piece of chromium steel with approx. 1 % C and 1.5 % Cr. Longitudinal section. Etch: Picral

ing with covers, for instance, iron rods are used, from which pieces may melt off and fall into the liquid steel. But more probably the metallic inclusions originate in floated base plates of carbon-deficient steel, such as are inserted into the mold in casting from above for the protection of the bottom. In any case, irrespective of their origin, they must be regarded as casting defects.

In the following examples metallic inclusions in steels of various types are shown.

1. The Figs. 1 and 2 show the structure of an inclusion in an annealed chromium steel with approx. 1 % C and 1.5 % Cr. It consists of

ferrite with lamellar pearlite. The carbon content of the inclusion was therefore considerably lower than that of the chromium steel and was adapted to the latter by diffusion only at the periphery of the inclusion. The lamellar formation of the pearlite shows that the inclusion has reached the $(\alpha + \gamma)$ — region during annealing of the chromium steel that took place between 750 and 800°C according to specifications. The structure of the chromium steel consisted of spheroidal carbide in a ferritic matrix. The carbon has gone into solution and transformed into carbide grains.

2. The Figs. 3 and 4 show the structure of a

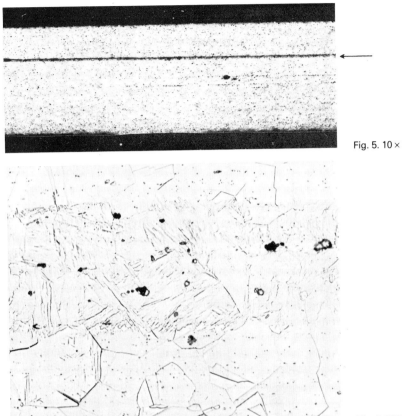

Fig. 5. 10 ×

Fig. 6. 500 ×

Figs. 5 and 6. Foreign inclusion in a pipe of 18/8 stainless steel. Longitudinal section through the weakly magnetized region, Etch: V2A pickle

section of a hardened piece of the same chromium steel that also contains metallic inclusions. While the chromium steel in this case had a structure of martensite with hypereutectic carbide, the inclusions consisted of a very fine laminated eutectoid of the lower pearlite range (Troostite).

The carbon content in this structure cannot be estimated. But it was lower than that of the chromium steel, because the latter was clearly decarburized in the region directly adjoining the inclusion. This is expressed in Fig. 4 where the hypereutectic carbide has disappeared from the structure of the chromium steel and the austenite has been transformed in some areas to Bainite. The inclusion itself has been transformed to pearlite as previously stated which proves that it is more alloy-deficient than the surrounding steel.

3. In a pipe of austenitic 18/8 stainless steel a weakly magnetizable spot of limited size was found. It was assumed that in this region an enrichment of δ-ferrite may be present.

X-ray diagrams of the outer and inner pipe surface in the indicated area showed only austenitic lines.

A longitudinal section through the weak magnetic region of the pipe wall showed the presence of a thin layer with deviating structure. This layer appeared as a dark strip at slight magnification (designated in Fig. 5 by arrow). During magnetic particle testing iron filings were deposited in this strip which proves that the weak magnetic reaction of the pipe was caused by this strip. During higher magnification it could be seen that the deviating etch behavior was caused by carbon-deficient martensite contents (Fig. 6). Therefore this inclusion too was probably more alloy-deficient than the austenitic steel, similar to the ones described above. All three cases therefore were casting defects of the type mentioned above.

[1] F. K. NAUMANN, F. SPIES, Prakt. Metallographie 4 (1967) 371/372

[2] F. K. NAUMANN, F. SPIES, Prakt. Metallographie 4 (1967) 541/546

[3] F. K. NAUMANN, F. SPIES, Prakt. Metallographie 9 (1972) 661/665

[4] F. K. NAUMANN, F. SPIES, Prakt. Metallographie 11 (1974) 43

Fracture of Tempered Leaf Springs

Friedrich Karl Naumann and Ferdinand Spies

Max-Planck-Institut für Eisenforschung

Düsseldorf

U-shaped leaf springs, intended to serve as spacers between oil tank floats and the inner walls of the containers, broke while being fitted, or after a short time in use, in the bend of the U. The springs were made of tempered strip steel of type C 88 with 0.84 % C, bent at room temperature and electroplated with cadmium for protection against corrosion.

Each fracture showed seven or eight kidney-shaped cracks (Fig. 2). At the origins of these cracks on the concave inner surface of the springs, crater-like depressions and beads of melted and resolidified material were found (Fig. 1). Such beads were also present elsewhere on the surface, and in some cases had already led to the formation of stress cracks in the high strength material (Fig. 3).

Fig. 1. Surface with melted droplets and craters at the fracture. 5 x

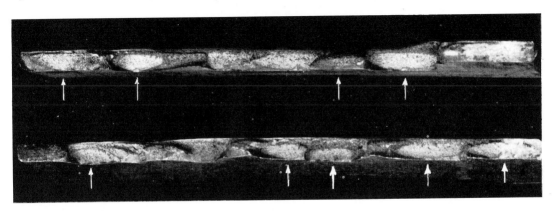

Fig. 2. Fracture with cracks (points of origin denoted by arrows). 2.5 x

A metallographic investigation showed that the material under the beads and craters at the origins of the fractures had been heated to the melting point and tranformed to a martensitic structure (Fig. 4). The transformed structure had a Vickers hardness (HV 0.2) of 733 to 773, compared with values of 447 to 463 for the unchanged tempered structure (cf. the hardness impressions in Fig. 4).

These observations indicate that the fracture of the springs was caused by stress cracks as a consequence of local hardening. The beads of melted material could have been produced by flying sparks from nearby welding operations, or by arcing across a bad contact between the springs and the current supply during electroplating. Experiments showed that in the former case melting

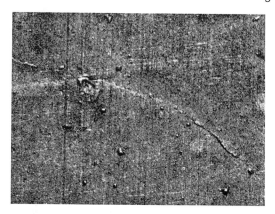

Fig. 3. Surface with melted droplets and stress crack. 25 x

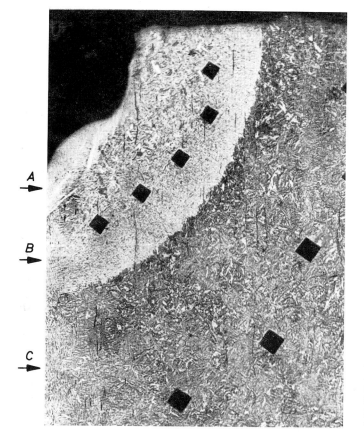

A →

B →

C →

Fig. 4. Microstructure in a cross-section through the origin of a crack, etched in alcoholic picric acid. 200 x
Zone A = melted
Zone B = transformed, ◇ HV 0.2: 733 to 773 kg/mm²
Zone C = unaffected, ◇ HV 0.2: 447 to 463 kg/mm²

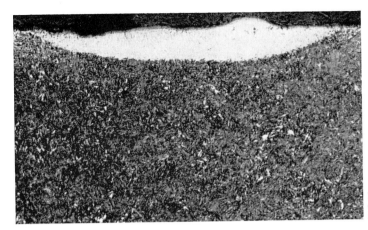

Fig. 5. Local melting and hardening caused by an electrical engraving tool, etched in alcoholic picric acid. 200 x

phenomena of the observed type do not arise, whereas local melting and martensite formation can be induced by briefly touching the surface with the point of an electrical engraving tool at a potential of 4 V (Fig. 5).

The hardening phenomena, and therefore the cracks, in the springs, would therefore appear to be due to a faulty procedure during electroplating with cadmium.

Steel Wire Cracked at Welded Joint

Friedrich Karl Naumann and Ferdinand Spies

Max-Planck-Institut für Eisenforschung
Düsseldorf

A steel wire of 2.3 mm diameter broke during cable twisting. The fracture occurred obliquely to the longitudinal axis of the wire and showed a constriction at the end. Therefore it was a ductile fracture. File mark type work defects were noticeable on the wire surface at both sides of the fracture, but they had no effect on the breakage of the wire.

Longitudinal sections were made for examination at the fracture and on each side of it at approximately 200 mm distance. Away from the fracture area the wire had a normal

Fig. 1. Normal patented structure. HV 5 = 490 kp/mm²

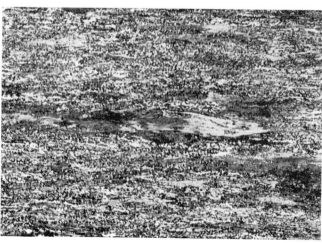

Fig. 2. 7 mm distance from fracture. HV 5 = 396 kp/mm²

Fig. 3. 3 mm distance from fracture.
HV 5 = 391 kp/mm²

Figs. 1 to 3. Structure of wire. Longitudinal section, etch: Picral. 500 x

structure of hyperfine lamellar pearlite (sorbite) of a "patented" and cold drawn steel wire (Fig. 1). In the vicinity of the fracture (Fig. 2) the cementite of the pearlite was partially spheroidized, while at the fracture itself it was completely spheroidized. Therefore the wire was locally annealed at this point. This could be confirmed by hardness measurements. Hardness (HV 5) had decreased from 490 to 390 kp/mm² at this site.

Apparently this was due to an annealed zone of the weld joint. This could also be confirmed in addition to the above mentioned "file marks" by individual surface decarburized folds that were noticeable in the zone, while the wire otherwise had a completely smooth surface.

It is likely that the wire cracked at this point during the last drawing and then broke during twisting due to its lower strength in the weakened cross section after prior deformation.

Fracture of a Bone Drill

Friedrich Karl Naumann and Ferdinand Spies

Max-Planck-Institut für Eisenforschung
Düsseldorf

Complicated bone fractures can be internally splinted by driving a stainless steel pin into the marrow cavity of the broken bone. To this end, the bone must be bored out in several stages. This is done with drills of various diameters, made of stainless tool steel, driven by a flexible coupling (Fig. 1).

One of these drills broke during an operation on a patient and was examined. It showed two fatigue fractures, one of which had started from a sharp-edged, coarsely milled slot (fracture A 1 in Fig. 2), and the other from a point on the outer sheath surface which was not subjected to particularly high stresses (A 2 in Fig. 2).

The fatigue fracture A 1 is easily accounted for by the stress concentration built up at this point as a result of the sharp edges and the coarse machining grooves. Fractures of this type were later found on other drills as well (Fig. 1). The relationship to the machining grooves can be seen particularly clearly in Fig. 3, which shows four fatigue breakages lying close to one another, as revealed after the crack in the slot had been broken open. On the other hand, an explanation for the fatigue fracture A 2 had to be sought. The discovery of the remains of a number, which had been inscribed with an electrical engraving tool for identification purposes, at the point of origin of the fracture, revealed the

Fig. 1. Bone drill with crack in the coupling. 3 x

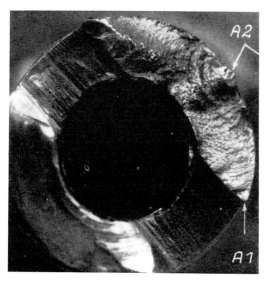

Fig. 2. Fracture surface with to fatigue fractures (A 1 and A 2). 10 x

Fig. 3. Crack in drill of Fig. 1, after being broken open, showing several fatigue fractures which had started from machining grooves. 15 x

cause (Fig. 4). It may have been the curved part of a 2 or 3. Next to this, on a section which is not so subject to fracture, a figure 5 can still be seen. The examination of a cross-section through this point confirmed that the material had been heated to the melting point during the engraving of the number, and that multiple cracking had occurred during cooling (Fig. 5). One of these cracks had certainly led to the development of the fatigue fracture A 2.

Electrical engraving tools of this type are often used by metallographers for marking their specimens. The metallographer should bear in mind that with this technique he can produce local changes in the microstructure through hardening or annealing, particularly in hardened specimens, not only at the point of the engraver but also, if the specimen is badly earthed, on the reverse side.

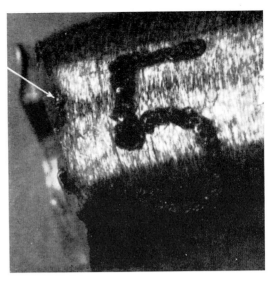

Fig. 4. Surface at the point of breakage. 10 x

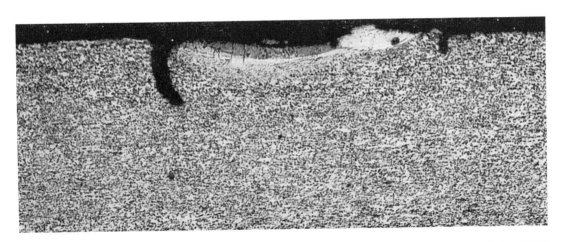

Fig. 5. Microstructure at the origin of the fracture A 2. Longitudinal section, etched in V2A pickling solution (20° C). 250 x

Broken Slide of a Friction Press

Friedrich Karl Naumann and Ferdinand Spies
Max-Planck-Institut für Eisenforschung
Düsseldorf

A short fracture section of a forged and normalized Ck 35 steel slide was to be analyzed for cause of failure.

The fracture is shown in Fig. 1. Three clearly distinct zones are readily evident, a dark colored crystalline incipient crack (right top) propagating into a far advanced, rubbed fracture surface, and a fine crystalline final break (left bottom).

Chemical analysis showed that the steel composition corresponded to the Ck 35 specification according to DIN 17 200. On the basis of the Brinell Hardness, found to be 153 to 154 kg/mm², the strength requirement of the standard for annealed Ck 35 of 50 to 60 kg/mm² was also fulfilled.

Metallographic examination showed that the steel was clean and free of defects. In order to analyze the cause of failure, it remained to be determined whether the dark incipient crack from which the fatigue crack propagated was already produced during the manufacture of the slide prior to annealing, or was formed during operating, e. g. through impact stresses. In the former case, the incipient crack had to be oxidized and decarburized, whereas it could be stained by oil residue, dirt or corrosion if formed subsequently. A longitudinal section L — — L through the fracture path showed a light case in the dark incipient crack region (Fig. 2) after macroetching. Microscopic examination disclosed that the structure adjacent to the crack consists of pure ferrite (Fig. 3). Therefore, the dark incipient crack was present before the last heat treatment, and was oxidized and decarburized prior to the conclusion of the annealing process.

The failure cause must therefore be sought in the manufacturing process of the slide.

Since the crack runs perpendicular to the fiber, it cannot have been formed before or during forging. It is probably a thermal stress crack which was produced during the flame

Fig. 1. Fracture. Approx. 0.5 x

cutting of the middle section of the slide. The initial crack acted as a sharp notch favoring the formation of the fatigue fracture which lead to the failure of the slide.

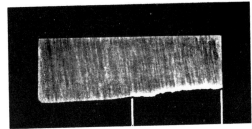

dark crack region

Fig. 2. Longitudinal section L — — L in Fig. 1, etching: copper ammonium chloride. 1 x ▶

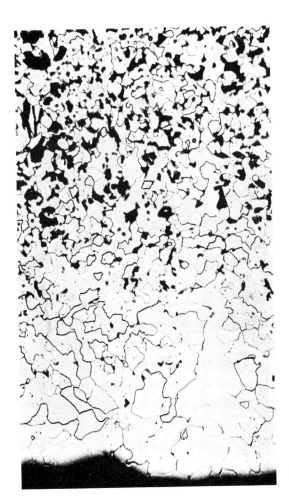

Fig. 3. Decarburized crack boundary (light zone in Fig. 2), etching: nital. 100 x

Worn Gears for Fuel Injection Pumps

Friedrich Karl Naumann and Ferdinand Spies

Max-Planck-Institut für Eisenforschung

Düsseldorf

Two fuel injection pump gears that were nitrided in a cyanide bath were submitted by the engine manufacturer for examination of hardness distribution and failure analysis. The gears showed signs of wear after only comparatively brief operation. They were made of the unalloyed steel C 45 (Material No. 1.0503) according to DIN 17 200 and were normalized.

Gear 1 with 1905 hours of operation showed at one side pittings on both flanks of the teeth as well as incipient fractures (Figs. 1 and 2). Gear 2 with 1713 hours of operation also showed at one side incipient fractures of the nitride layers at the outer part of the teeth (Fig. 3).

The ferritic-pearlitic grain structure of the gears (Fig. 4) corresponded to that of a normalized C 45 steel. The compound zone of

Fig. 1

Fig. 2

Figs. 1 and 2. Failed gear 1. 5 ×.

Fig. 3. Failed gear 2. 5×.

Fig. 4. 200×.

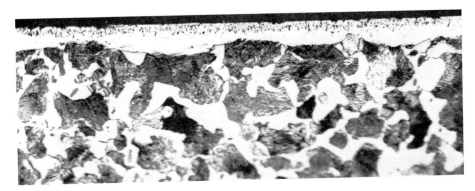

Fig. 5. 500×.

Figs. 4 and 5. Peripheral structure of a non-worn area. Transverse section, etch: Picral.

ε-nitride under the surface was 15 to 20 μm thick and was porous in the outer part, as is usual (Fig. 5). At the flanks of the teeth it was partially fragmented under operational stress. The micrograph showed the beginnings of pitting (Fig. 6). The depth of the diffusion zone could not be exactly established metallographically.

Hardness tests with a 25 p testing load (not absolute values) of a transverse section below the tooth back of the failed gears as well as of an unfailed comparison gear (3) resulted in the curve plotted in Fig. 7. The nitride zone values were 525 to 780 kp/mm². The wide scatter can be explained by porosity and low adhesive strength. The diffusion zone showed

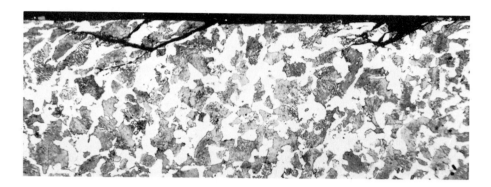

Fig. 6. Pitting of a tooth flank. Transverse section, etch: Picral. 200 × .

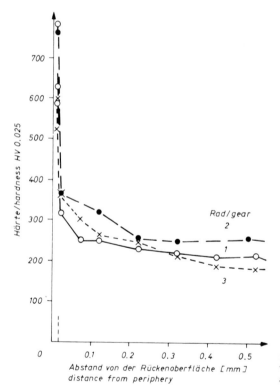

Fig. 7. Hardness distribution from periphery to core on transverse sections of non-worn areas of gears 1 and 2, and of the unused gear 3.

a hardness of 320 to 360 kp/mm^2 at the boundary to the transition layer. It had a depth of 300 μm where hardening was apparent.

Such a nitride layer can increase the wear resistance and fatigue strength in cases of main friction. But it did not stand up to the high and one-sided compressive stress applied in this case and could not prevent pitting. It could even have accelerated the wear by the incipient break down of the nitrided layer. Gas nitriding at greater depth under application of a suitable special steel or case hardening would have been better under these circumstances.

Stress Cracks in Brass Pipe Couplings

Karin Kuhn
Max-Planck-Institut für Metallforschung
Institut für Sondermetalle
Stuttgart

The brass pipe couplings submitted for examination were deep-drawn from disks then annealed and subsequently cold threaded. Chemical analysis confirmed that the specified alloy Ms 63 was used for fabrication.

Some of the pipes already showed fine cracks prior to their installation (Fig. 1); in most cases however the cracks were detected after a certain period of operation.

Metallographic examination of sections

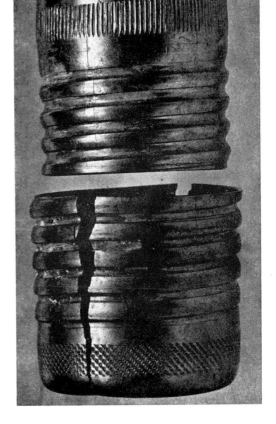

Fig. 1. Pipe coupling with cracks before installation. 1 x

Fig. 2. Pipe coupling cracked during operation. 1 x

Fig. 3. Longitudinal polished section through threads of a pipe coupling before installation. 7 x

Fig. 4. Structure of the region of heavy deformation (location a in Fig. 3). 100 x

through the threads (Fig. 3) reveals the changes in shape produced by the process of forming the threads by rolling. An increase in deformation is noted particularly at the right side of the thread (Fig. 3, location a). The typical deformation structure of such a region can be recognized in Fig. 4 by the elongated grains and by the large number of slip-lines. In contrast, the region between the threads (Fig. 3, location b) only shows slight deformation as evidenced by the scarcely elongated grains and small number of slip-lines (Fig. 5).

Figure 6 shows the microstructure of a polished cross-section through the threads. The intercrystalline course of the cracks indicates stress-cracking as it often appears in brass after heavier cold deformation.

Therefore a test was conducted according to DIN 1785 to check for sensitivity to stress cracking. This consists in immersing a brass specimen in a 1.5 % mercuric nitrate solution for 15 min. If after this time no cracks appear, the specimen is considered stress-free.

One of the specimens which, even under more careful observation, did not manifest any cracks in the as-received condition, was left in the above solution for 15 min. The cracks seen in Fig. 7, which confirmed the suspicion of stress-cracking, developed after only 3 minutes' immersion.

The splitting of the couplings could have been avoided by a tempering heat treatment at temperatures between 230 and 300° C after rolling the threads. This procedure will on one hand reduce the internal stresses; on the other hand the strengthening gained by the cold deformation is maintained.

Fig. 5. Structure of region between threads (location b in Fig. 3). 100 x

Fig. 6. Course of crack in microstructure; etchant: copper ammonium chloride – ammonium hydroxide. 100 x

Fig. 7. Pipe coupling cracked after immersion in 1.5 % mercuric nitrate solution. 0.8 x

Broken Inner Rings of Spherical Roller Bearings

Friedrich Karl Naumann and Ferdinand Spies
Max-Planck-Institut für Eisenforschung
Düsseldorf

Inner rings of spherical roller bearings out of the full hardening ball bearing steel 100 CrMn 6 with about 1 % C, 1.1 % Mn and 1.5 % Cr (Material No. 1.3520) failed in service. Due to the cracks, parts from the middle flange broke (Fig. 1) or the rings failed in radial direction completely (Fig. 2). All the cracks and fracture originated from the middle flange (Fig. 3). In all of the three rings one flank showed heavy wearing and scouring (Fig. 4). The cracks started from the edge of this flank with the cylindrical mantle surface of the middle flange. This can be seen in Fig. 5, which shows a fractured crack. The cracking resembles a fatigue cracking, but in a fine-grained hardened steel like this the fracture-faces due to stress-cracking and overload fracture look the same as this. The forced rest fracture has a satin-like fine structure, which proves a fault-free hardening.

The metallographic examination of microsections showed that one flank of the middle flange, the same from which the cracks originated, had zones of changed structures

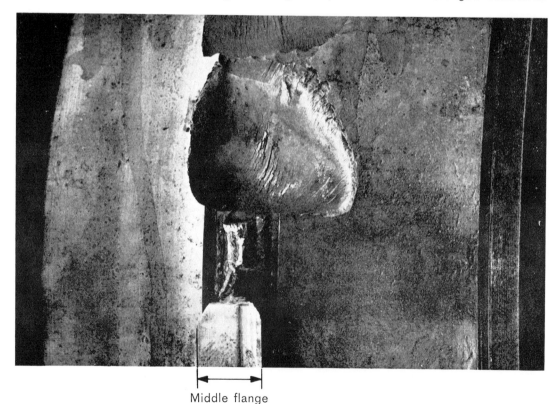

Middle flange

Fig. 1. Inner ring with breakouts from middle flange. 0.8 x

Fig. 2. Fracture of an inner ring. 0.8 x

Middle flange

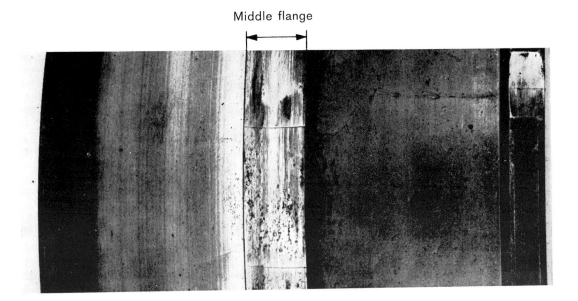

Fig. 3. Bearing surface of the ring with cracks in middle flange. 0.8 x

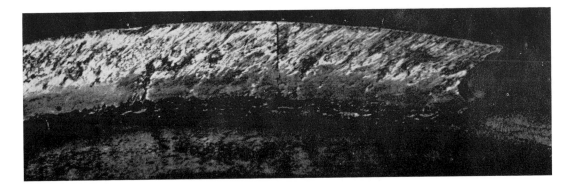

Fig. 4. Worn flank of the middle flange. 2 x

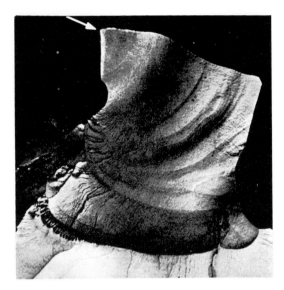

Fig. 5. Opened crack in middle flange. Crack origin — see arrow. 2 x

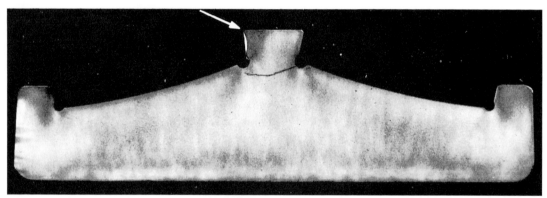

Fig. 6. Radial microsection through ring with crack in middle flange. Arrow = worn flank with structure due to hardening. Etching: Nital. 0.8 x

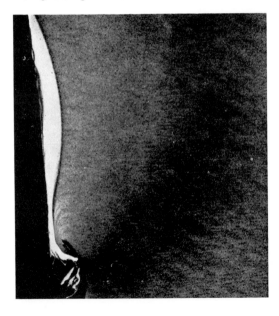

Fig. 7. Worn flank from Fig. 6. 8 x

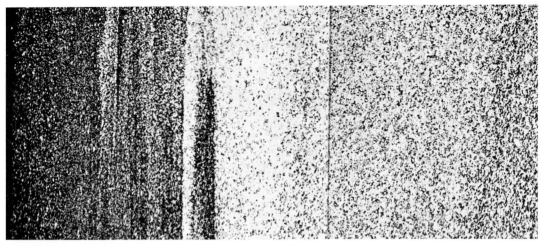

▲
Fig. 8. Structure below worn flank (Fig. 7). Cross-section, etching: Nital. 500 x

Fig. 9. Structure of core of the rings. Etching: Nital. 500 x

(Figs. 6 and 7). This flank, as the microsections show, is so far worn down that the edge which was bevelled originally has now a knife-edge. The structure of the edge of the worn flank shows a crescent shaped zone of fine structured martensite in front of a broken, narrow and dark annealed zone with carbide precipitation (Fig. 8). This shows that this flank had been heated up on the surface to a temperature reaching into austenite zone and then cooled very fast with a following annealing. The temperature was so high that even overeutectic carbides have partly dissolved. Adjoining this hardened zone is an annealed zone, which is so soft, that the material with the martensitic case has deformed under the pressure of the rolls into the rounded off groove (Fig. 7). The microstructure goes over a tempered zone to the microstructure of the core (Fig. 9), which consists of fine-needled, low-annealed martensite with embedded regularly distributed fine carbides, which reinstates the conclusion drawn from the fracture surface that the heat-treatment was done properly. The failure of the rings is therefore a result of repeated heating and rapid cooling of the surface due to the grinding of the bearings on one flank of the middle flange. The stress-cracks ("grind-cracks") [1] originating herefrom have apparently spread in steps which finally led to the breaking off of parts from the middle flange and complete failure of the rings.

[1] F. K. NAUMANN, F. SPIES, Prakt. Metallographie 5 (1968) 291/298

Fractured Three-Cylinder Crankshaft

Friedrich Karl Naumann and Ferdinand Spies

Max-Planck-Institut für Eisenforschung
Düsseldorf

The fracture cause had to be determined in a three-cylinder crankshaft made of chrome steel 34 Cr 4 (Material No. 1.7033) according to DIN 17200. The fracture had occurred after only 150 hours of operation. The fracture is of the bend fatigue type which originated in the fillet of the main bearing and ran across the jaw almost to the opposite fillet of the adjoining connecting rod bearing (Fig. 1). The fillet is well rounded and smoothly machined. Thus, no reason for the fracture of the crankshaft could be found externally.

A section was cut longitudinally through the fracture origin of the pins in order to establish whether a material defect might have caused the damage. Figure 2 shows this section after etching with copper ammonium chloride solution according to Heyn. No material defects were discernible either in the origin or anywhere else. The fiber path is flawless. The bearing surfaces are case hardened to a depth of approximately 2.5 mm. A sulfur print according to Baumann showed the same result.

The crank contained thin longitudinally extending manganese sulfide inclusions in the usual amount characteristic of manganese-alloyed steel. The purity of the material may be characterized by group M 3 of the

Fig. 1. Fracture piece with main bearing. 1 x

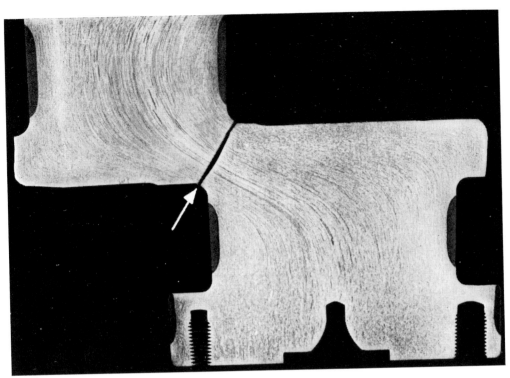

Fig. 2. Longitudinal section through crack path (designated by arrow), etch according to Heyn. ³/₄ x

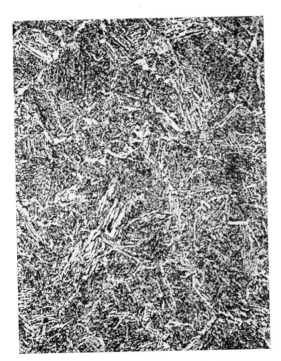

Fig. 3. Core structure of the crank. Longitudinal section. Etch: Picral. 500 x

Steel-Iron-Testsheet 1570. The core structure (Fig. 3) is almost free of ferrite, meaning that the crank is well heat-treated.

The chemical composition corresponded to the specification for 34 Cr 4 as could be verified by analysis. Core hardness (HB 5/750) however, was rather low at 230 kp/mm² corresponding to a tensile strength of 78 kp/mm².

Accordingly, no cause for the crank fracture could be established from the material testing. Probably the load was too high for the strength of the crank. Tensile strength could have been increased for the same material by tempering at lower temperature. Additionally the resistance against high bend fatigue stresses or torsion fatigue stresses could have been substantially increased by including the fillet in the case hardening process.

Failure in Steam Turbine Blades

Franc Vodopivec, Ladislav Kosec, Roman Brifah and Bogomir Wolf
Metallurgisches Institut
Ljubljana, Jugoslavia

1. Description of damage and mechanical tests on the steel

When a steam turbine was put out of service, cracks were noticed on many of the blades in the low pressure section round the stabilisation bolts and perpendicular to the blade axis (Fig. 1). According to the literature [1]) such blades are made from chrome alloy steel X 20 Cr 13 (material no. 1.402). According to the Iron and Steel specification Sheet 400 this steel should contain 0.17 to 0.22 wt. % C and 12 to 14 wt. % Cr. Chemical analysis of the blades yielded 0.20 wt. % C and 13.3 wt. % Cr. The mechanical properties are given in Table 1. They correspond to specification.

The surface hardness was measured on two of the blades. This experiment revealed that the hardness round the bolts in blades with or without cracks was significantly higher than over the rest of the surface (Fig. 2), in some places more than 100 %.

Etching revealed that the region of greater hardness possessed a different macrostructure from the normal regions of the blade. This different macrostructure appeared round all the bolts at approximately the same place and differed only size.

2. Metallographic and microfractographic investigations

Specimens for microscopic examination were taken at various characteristic places from damaged and unused reserve blades. It was discovered that the steel contained few in-

Table 1. Mechanical properties of the turbine blades under investigation

	Mean of 3 blades	Specification for X 20 Cr 13
Breaking strength (kp/mm²)	76,8	65–30
Yield point (kp/mm²)	60,8	min. 45
Strain δ 5	18	min. 16
Necking (%)	64	–
Notch toughness (kpm/cm²)	7,6	min. 4
Hardness (kp/mm²)	227	180–250

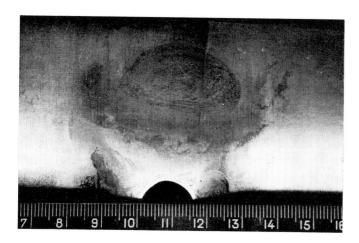

Fig. 1. Cracks round the bolt in a turbine blade. Macroscopically etched with FeCl₃ + HCl + ethanol. 0.7 x

clusions and consisted of tempered martensite which was somewhat finer in thin cross sections than in thicker (Figs. 3 a and b). The microstructure of the harder regions round the bolts consisted of fine grained untempered martensite with a more or less well developed carbide network at the grain boundaries (Figs. 4 a and b). The main crack and its lateral branches followed the grain boundaries. Such a microstructure can arise only if the steel is heated to the austenite region, i. e. from 850 to 900 °C, and cooled slowly through the three phase $\alpha + \gamma +$ carbide region.

When the bolts were brazed into the blades inadmissible localised overheating of the steel must have occurred, which resulted in transformation stresses and hence reduced deformability. Internal tensile stresses were detected using strain gauges in the brazed uncracked blade at the edge of the temperature affected zone. The stress was 14.3 kgf/mm². It was greatest approximately in the direction of the blade axis. In a new blade without bolts the greatest stress at the same place and in the same direction is only 2.9 kgf/mm². It can be concluded that the stresses in the microscopic volume were even greater since volume changes due to localised heating, cooling and transformation can give rise to internal stresses which can exceed the yield point of the steel, or as evidenced by the production of hardening cracks, even the breaking strength. This indicates that the cracks arose as a consequence of careless brazing.

The question arises as to whether the cracks should be considered as stress cracks over their entire extent or partially as fatigue

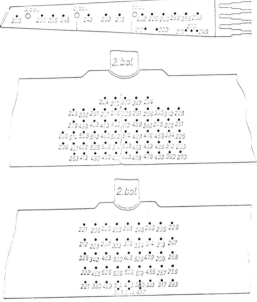

Fig. 2 top: hardness (HV 10) along the blade axis (1:3.6), centre and bottom: hardness measured at the bolts with or without cracks on a quadratic grid (1:1)

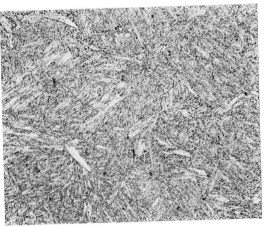

Fig. 3 a. Microstructure of steel at the thinner cross section

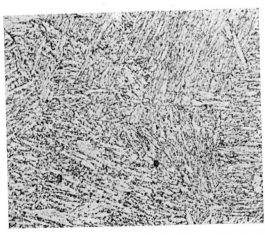

Fig. 3 b. Microstructure of steel at the thicker cross section

Figs. 3 a and b. Cross section etched with $FeCl_3$ + HCl + ethanol. 500 x

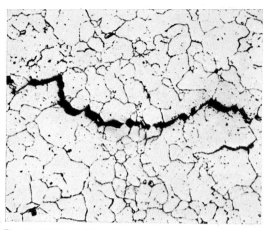

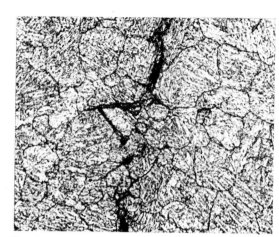

Fig. 4 a. Etched with sodium picrate

Fig. 4 b. Etched as in Fig. 3

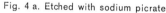

Figs. 4 a and b. Microstructure in the vicinity of the brazed zone, cross section. 200 x

cracks produced by vibration in the operation of the turbine as a result of steplike growing of microcracks. This could not be deduced from the fracture surfaces. A microfracto- graphic investigation was therefore carried out. The fracture surface was cleaned by pressing into plastic Rodoid from which indirect carbon replicas were then made. Observation in the electron microscope showed a topological periodicity at individual points on the fracture surface (Fig. 5). This is a characteristic indication that the crack developed in stages.

3. Conclusions

The following conclusions can be drawn on the basis of the experimental results: The inappropriate heating during the brazing of the bolts led to localised damage to the steel and caused internal stresses and embrittlement. Where the stresses were greatest, microcracks appeared on cooling after brazing or as a result of the vibration of the turbine. The cracks propagated stepwise by a mechanism typical of the propagation of fatigue cracks.

The causes of failure cannot be removed by subsequent treatment of the steel. A brazing alloy with a low melting point must be used in future.

Fig. 5. Microfractography of the fracture surface. Two stage Rodoid-carbon replica. 6000 x

Literatur/References

1) F. RAPATZ, Die Edelstähle, Springer Verlag, Berlin, Heidelberg (1962)

Boiler Tube Cracked During Bending

Friedrich Karl Naumann and Ferdinand Spies

Max-Planck-Institut für Eisenforschung

Düsseldorf

A seamless hot-drawn boiler tube NW 300 of 318 mm O.D. and 9 mm wall thickness made of steel 15 Mo 3 was bend with sand filling after preheating allegedly to 1000° C. In the process it had cracked repeatedly in the drawn fiber (Fig. 1).

A check of the chemical composition showed the following values:

C %/o	Si %/o	Mn %/o	P %/o
0,13	0,17	0,53	0,032

	S %/o	Mo %/o	Cu %/o
	0,022	0,26	0,26

The composition thus corresponded to specifications. Only the exceptionally high copper content was noticeable.

For metallographic examination one longitudinal section was taken out of the straight part and out of the tube bend, respectively. The initial microstructure of the straight part consisted of a fine-grained mixture of ferrite and pearlite (Fig. 2), whereas the structure of the bend consisted of a very coarse-grained Bainite with little ferrite (Fig. 3).

Accordingly, the tube was strongly overheated during bending. Furthermore, it also distinctly showed signs of burning. The surface was strongly scaled. Under the scale there appeared additionally precipitates of metallic copper next to oxidic precipitates ("internal oxidation"). This copper had accu-

Fig. 1. Transverse cracks on external tube bend. 1 x

mulated in the residual iron as difficult-to-oxidize metal (Figs. 4 and 5). The oxidation had penetrated already 3 mm into the steel along the grain boundaries (Fig. 6).

Therefore the damage is due to overheating and burning during preheating and bending. Furthermore, crack formation was promoted by precipitation of metallic copper that had penetrated into the austenitic grain boundaries under the influence of tensile stresses that arose during bending. This phenomenon is known under the designation "solder brittleness"[1].

[1] vgl. ED. HOUDREMONT; Handbuch der Sonderstahlkunde, 3. verbesserte Aufl., Bd. 2, S. 1273

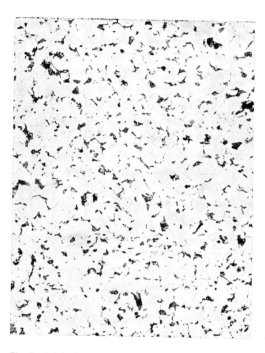

Fig. 2. Original microstructure in straight part

Figs. 2 and 3. Etch: Nital. 200 x

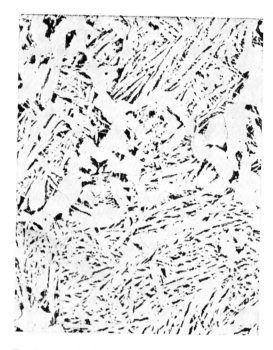

Fig. 3. Microstructure of bend with cracks

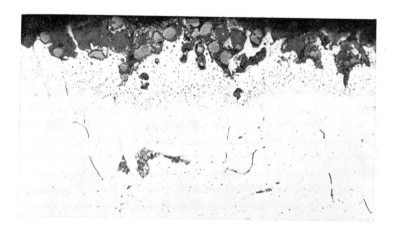

Fig. 4. 500 x

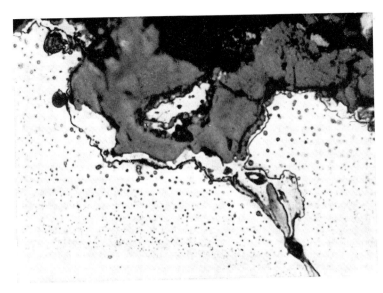

Fig. 5. 1000 x

Figs. 4 and 5. Oxidized edge, etch: Nital

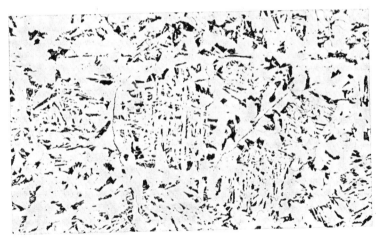

Fig. 6. Oxide precipitates at austenitic grain boundaries 3 mm under surface, etch: Nital. 200 x

Broken Connecting Rod from a Motor Boat

Egon Kauczor

Staatliches Materialprüfungsamt an der Fachhochschule
Hamburg

The connecting rod was broken in two places at the small end. The general view of the reassembled pieces in Fig. 1 shows the location of the fracture surfaces. At position I there is a fatigue fracture brought about by operational stress, whereas the fibrous fracture surface II is a secondary tensile fracture. Furthermore the transition on the other side of the rod is cracked symmetrically to the fatigue fracture (position III). Photographs of the broken off shell in Figure 2 show (top) the fatigue fracture at the transition between the rod and the small end and (bottom) the tensile fracture. Magnetic inspection showed up indications of cracking at the transition between rod and small end in six other connecting rods from the same batch.

Metallographic examinations of a connecting rod selected by a magnetic crack test revealed deep folds in the flash zone. As shown in Fig. 3 these folds are filled with scale and a broad zone around them is decarburized. The microstructure of the remaining material corresponds to correct annealing. Since the flash zone was ground after forging no decarburization of the surface can be detected outside the fold zone.

The results of the investigation show that the connecting rods were rendered susceptible to fatigue by the notch effect of coarse folds formed during forging.

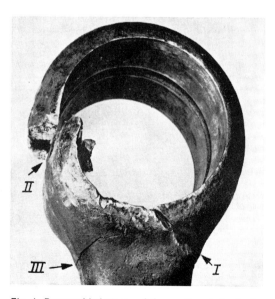

Fig. 1. Reassembled pieces of the small end of a broken connecting rod. ¼ x

Fig. 2. Fracture surfaces on the broken off shell. Top: Fatigue fracture surface (site I), Bottom: tensile fracture surface (site II). ¼ x

Fig. 3. Microsection through a fold in the flash zone of a connecting rod selected by the magnetic crack test. 50 x

Fracture of a Lifting Fork Arm

Günter Paul

Metal Mechanics Division
National Mechanical Engineering Research Institute, CSIR
Pretoria, South Africa

One arm of a lifting fork with an approximate cross-section of 150 x 240 mm fractured after only a short service life.

The fracture surface had the appearance of a forced fracture starting from a surface crack as shown in Fig. 1. After removing the paint from the surface of the arm a large number of surface cracks became visible. These cracks penetrated the material to a maximum depth of about 3 mm, and appeared to have originated during the forging of the fork. Paint which had penetrated into the cracks confirmed this conclusion. Forging at too low a temperature could have caused the cracks.

The results of a chemical analysis showed that the steel from which the fork was made was EN-25. This steel is suitable for this type of application and can be heat treated satisfactorily even in relatively large sizes.

The metallographic examination revealed a rather coarse bainitic structure as shown in Fig. 2, which is an indication that the re-sistance of the steel to withstand shock loads was low. Shock loads are unavoidable in the operation of a fork lifter. Impact tests (Charpy-V-Notch) on test specimens with square cross-section at room temperature, indicated an energy absorption of only 1.75 kpm/cm² confirming the expected low shock resistance.

An attempt was made to improve the toughness of the steel by a suitable heat treatment, consisting of quenching from 860° C in oil and subsequent tempering at 650° C. This treatment resulted in a considerably finer structure of tempered martensite as shown in Fig. 3. The improvement of the toughness by the described heat treatment was confirmed by impact tests, which yielded energy absorption values of 11.87 kpm/cm² representing a sevenfold improvement as compared with the original condition. The finer structure resulted additionally in a small improvement in the material's hardness.

In conclusion it can be said that the main cause of the present failure was the brittle-

Fig. 1. Fracture surface. ½ x

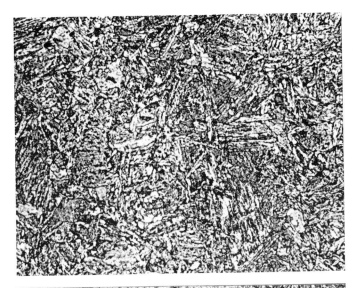

Fig. 2. Coarse bainitic structure of steel in condition as received, etched with Nital. 200 x

Fig. 3. Structure of the steel after the heat treatment (tempered martensite), etched with Nital. 200 x

ness of the material. The coarse structure (the reason for the low toughness of the steel) resulted from either insufficient or no heat treatment after the forging operation. The surface cracks assisted the initiation of fracture.

Cracked Cast Iron Crankcases

Friedrich Karl Naumann and Ferdinand Spies

Max-Planck-Institut für Eisenforschung

Düsseldorf

The front wall of a cast iron crankcase cracked at the transition from the comparatively minor wall thickness to the thick bosses for the drilling of the bolt holes. This transition was abruptly formed as can be seen from Figs. 1 and 2. The dark coloration of a part of the fracture plane indicated that the crack originated in the thin-walled part. This crack propagation led to the conclusion that shrinkage stresses were the cause of the cracks, because the adjoining thick- and thin-walled cast parts freeze and shrink at different times. A differential shrinkage causes the more damage, that is, the stresses at the cross sectional transition are the higher, the faster the piece cools down, and the more abrupt the cross sectional transition is formed. The stresses may already lead to crack formation during cooling, or they may remain in the piece as residual stresses, which are later superimposed on operating stresses and thereby favor the fracture at this particular point.

As could be seen from the metallographic examination the case was additionally aggrevated by the fact that the casting had a ferritic basic structure and the graphite in part showed a granular formation (Figs. 3 and 4), so that strength of the material was low.

In a second crankcase with the same crack formation the structure in the thick-walled part was better (Figs. 5a and b). But it also showed granular graphite in the ferritic matrix (Figs. 6a and b) in the thin-walled part

Fig. 1. Fragments of first crankcase. 1 x

Fig. 2. Fracture of first crankcase. 1 x

Fig. 3. Normal flake graphite

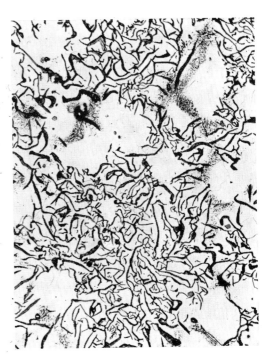

Fig. 4. Granular graphite

Figs. 3 and 4. Structure of first crankcase, etch: Picral. 200 x

Figs. 5 a and 6 a. Unetched

Figs. 5 b and 6 b. Picral

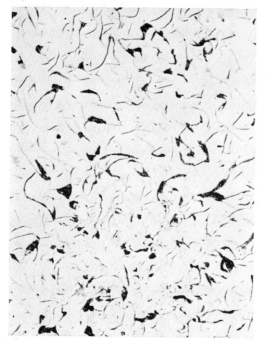

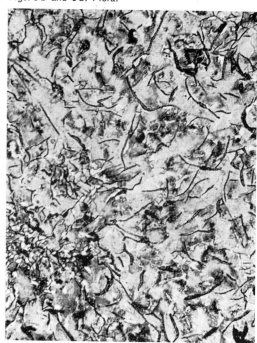

Figs. 5 a and b. In the thick-walled part

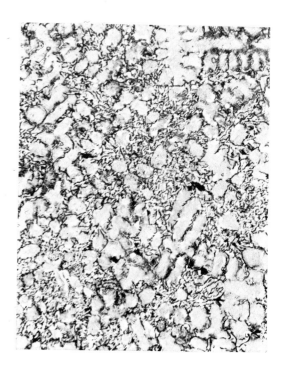

Figs. 6 a and b. In the thin-walled part with crack

Figs. 5 and 6. Structure of second crankcase. 100 x

between the dendrites of the primary solid solution precipitated in the residual melt.

A third crankcase (Fig. 7) had fractures in two places, first at the frontal end wall (→ a) and second at the thinnest point between two bore holes (→ b). Figures 8 and 9 show in larger scale sectors of the front wall with the fracture which is stained with alumina for easier recognition. In the cross sectional cut of Fig. 10 it can be discerned more easily that the fracture occurred at that point where the case is but 5 mm thick, whereas it becomes thicker toward both sides. The structure in this case as well as that of the thin-walled part of the second case consisted of pearlitic dendrites and a connecting enveloping network of ferrite and granular graphite (Fig. 11). The same structure could be found at the other point of fracture (b), where the wall between the borings in the raw casting was only 6 mm thick and was later worked down to 2.3 mm (Fig. 12). The thick-walled places on the other hand had a normal ferrite-deficient structure (Fig. 13).

In all three cases casting stresses caused by unfavorable construction and rapid cooling must be held responsible for the crack formation. Additional impetus was probably provided by the formation of a supercooled microstructure of low strength based on the same causes.

A fourth crankcase had cracked in the bore-

Fig. 7. Third crankcase with cracks. 0.25 x

Fig. 8. External view

Fig. 9. Internal view

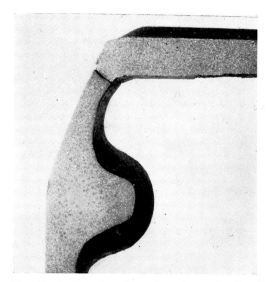

Fig. 10. Section through crack region a in Fig. 7, etch: Picral. 1 x

hole of the frontal face. One of the cracks was so narrow that it was made visible only by magnetic particle inspection (Figs. 14 and 15). Both branches could be discerned as independant cracks according to the fracture path.

The structure of the casting was permeated with clusters of granular graphite and ferrite near the surface as well as in the interior (Figs. 16 and 17).

In this case the cause of the fracture must be considered to be the low strength of a region that was caused by the bad microstructure and which was further weakened by the bore hole.

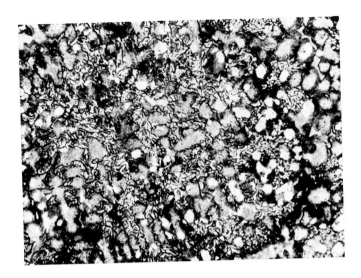

Fig. 11. In the thin-walled part at the crack region a in Fig. 7. 100 x

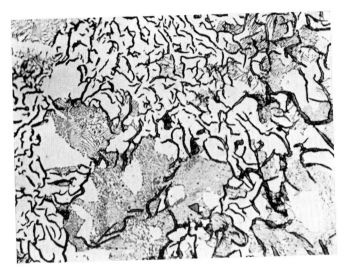

Fig. 12. In the thin-walled part at the crack region b in Fig. 7. 500 x

Fig. 13. In the thick-walled part next to crack region a in Fig. 7. 500 x

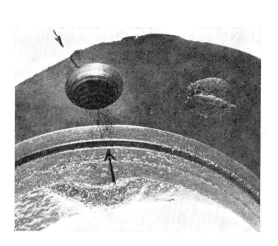

Fig. 14. End face

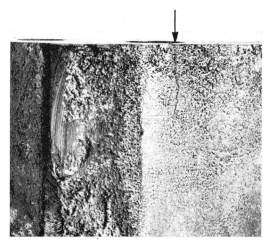

Fig. 15. External face

Figs. 14 and 15. Sections with crack region from the fourth crankcase after magnetic particle inspection. 1 x

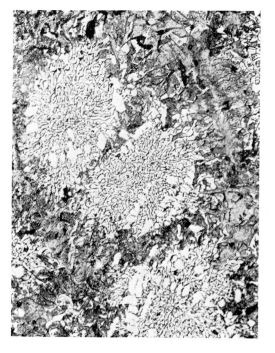

Fig. 16. Near surface

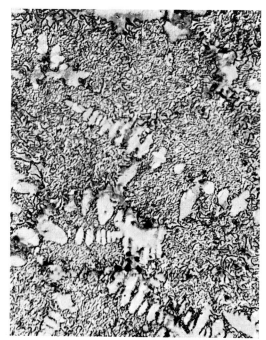

Fig. 17. Inside

Figs. 16 and 17. Microstructure of fourth crankcase, etch: Picral. 100 x

Cracked Bearing Caps Made of Cast Iron

Friedrich Karl Naumann and Ferdinand Spies

Max-Planck-Institut für Eisenforschung

Düsseldorf

Cast iron bearing caps in tractor engines fractured repeatedly after only short operating periods. Two of the pieces were examined initially. Both showed the same phenomena, externally as well as in metallographic section. The fracture originated in a cast-in groove and ran approximately radially to the shaft axis as shown in Fig. 1. The smallest cross section was at the point of fracture.

Furthermore, a shape-conditioned stress peak may have been formed at the bottom of the groove that had a rounded-off radius of only 2 mm.

For metallographic examination, sections were made at the notch base parallel to the fracture. Figure 2 shows one of the sections at a 3:1 scale. The larger ferrite clusters are

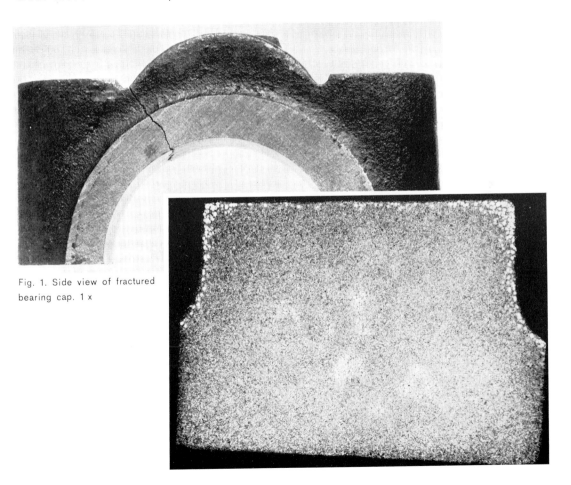

Fig. 1. Side view of fractured bearing cap. 1 x

Fig. 2. Section through groove parallel to fracture, etch: Picral. 3 x

especially noticeable. They are permeated by fine granular graphite which could be seen at higher magnification (Figs. 3 and 4). In one piece a narrow streak with ferritic matrix (Fig. 5) extended under the entire surface. This led to the conclusion that the casting had been annealed at a comparatively high temperature for stress relief. The core structure of the caps consisted of graphite in pearlitic-ferritic matrix (Figs. 6 and 7). The average Brinell hardness was 160 kgf/mm², from which a low strength could be inferred.

Accordingly, the following may be stated about the fracture cause: Casting stresses did not play a decisive role because of the simple shape of the pieces that were without substantial cross sectional variations. There appeared to be two factors exerting an

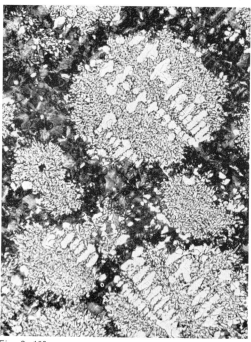

Fig. 3. 100 x

Fig. 4. 500 x

Figs. 3 and 4. Microstructure in edge zone, etch: Picral

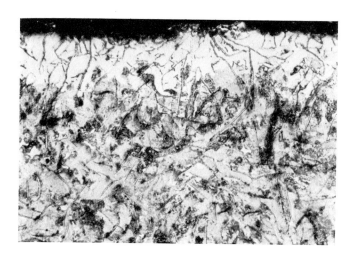

Fig. 5. Microstructure at surface, etch: Picral. 100 x

Fig. 6. 100 x

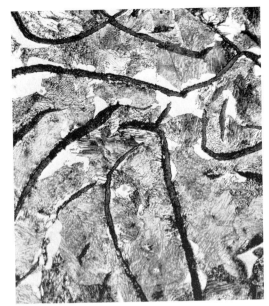

Fig. 7. 500 x

Figs. 6 and 7. Core structure, etch: Picral

unfavorable effect in addition to the comparatively low strength. First, the operating stress was raised locally by the sharp-edged groove, and second, the fracture resistance of the cast iron was lowered at this critical point by the existence of a ferritic bright border. The granular precipitate of the graphite and the ferritic microstructure caused thereby are a consequence of delayed solidification due to undercooling. The ferrite formation may have been further accelerated by the annealing. This structure showed low strength that had a particular damaging effect on the bottom of the groove at high operating stress.

In order to avoid such damage in the future it was recommended to observe one or more of the following precautions: 1. Elimination of the grooves, which would simultaneously strengthen the cross section and relieve the notch effect. 2. Removal of the ferritic bright border. 3. Slow cooling in the mold in order to avoid undercooling and therefore the formation of granular graphite. 4. Innoculation of finely powdered ferrosilicon into the melt for the same purpose. 5. Annealing at lower temperature or elimination of subsequent treatment in consideration of the uncomplicated shape of the castings.

Three more fractured bearing caps were subsequently examined in order to state particulars for this specific case about the wellknown effect of chemical c o m p o s i t i o n [1]). These were compared with 5 others that gave satisfactory service over longer operating periods. Mean values for both groups are given in the following table:

Bearing caps	C %	graphite %	Si %	Mn %	P %	S %	S_c
fractured	3,55	3,05	2,74	0,65	0,109	0,064	1,06
not fractured	3,25	2,73	2,62	0,66	0,312	0,124	0,97

In addition to the stated determinations, spectrographic analyses were conducted for additions of chromium, nickel, copper, vanadium, titanium among others, but in no case could additions be found that exceeded the customary quantities. The saturation value S_c that signifies the position of the alloy in relation to the eutectic composition ($S_c = 1$) was calculated according to the formula

$$S_c = \frac{C}{4.23 - 0.312 \cdot Si - 0.275 \cdot P}$$

This value is important because it is a mea-

sure of the tendency toward carbide decomposition as well as of the strength of the cast iron.

Carbon, graphite and silicon contents were significantly higher for the fractured caps than for the satisfactory caps, while phosphorus and sulphur contents were lower. Saturation value, too, was higher for the fractured caps than for the satisfactory ones in spite of lower phosphorus content. All unsatisfactory caps showed a hypereutectic composition with $S_c = 1.03$ to 1.09, while all satisfactory ones had a hypoeutectic compo-

Fig. 8. Cap of hypoeutectic cast iron. 3 x

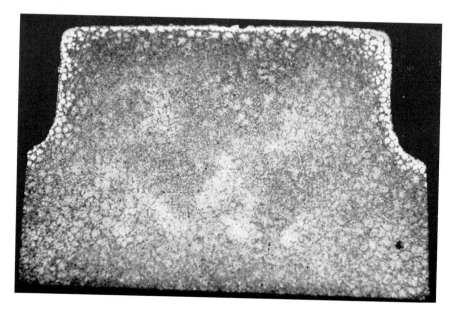

Fig. 9. Cap of hypereutectic cast iron. 3 x

sition with S_c = 0.96 to 0.98. According to the relation stated by P. A. H e l l e r and H. J u n g b l u t h [1]) σ_B = 100 — 80 S_c the fractured caps should show a strength of 15 kgf/mm², while the good caps should have a strength of 22 kgf/mm². This is expressed in the Brinell hardness as well, that has an average of 164 for the fractured caps and of 204 kgf/mm² for the satisfactory caps.

Accordingly, metallographic examination of sections of the groove bottom showed for the good caps an improved, i. e. a ferrite-deficient edge and core structure (Figs. 8 and 10) as compared to the broken ones (Figs. 9 and 11).

The composition of the cast iron should therefore be held in the hypoeutectic alloy range.

It remained to be investigated what effect s t r e s s r e l i e f a n n e a l i n g had on the ferrite content of the structure. For this purpose a number of untreated caps were annealed for 2 h at 550, 575, 600, 650 and 750° C, respectively, after their structure was determined in the as-cast condition.

Even the non-annealed specimens showed ferrite clusters with granular graphite under the surface as well as in the core in minor proportion corresponding to the favorable composition of the casting. Therefore the ferrite did not result from carbide decomposition during annealing. No change was noticeable in the structure of the annealed spe-

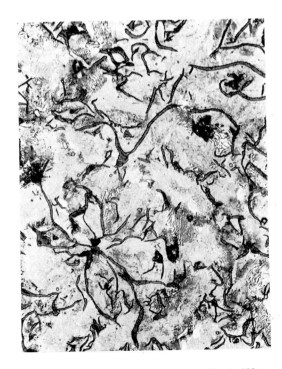

Fig. 10. Cap core structure according to Fig. 8. 200 x

Fig. 11. Cap core structure according to Fig. 9. 200 x

Figs. 8 to 11. Effect of composition on microstructure, radial sections in groove, etch: Picral

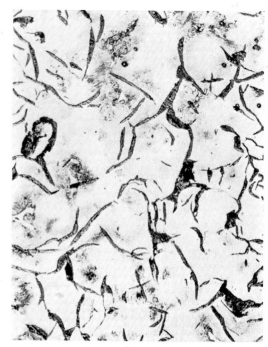

Fig. 12. 2 h at 625°/air. 100 x

Fig. 13. 2 h at 650°/air. 200 x

Figs. 12 and 13. Effect of annealing temperature on core structure of bearing caps of hypoeutectic cast iron, etch: Picral

cimens at temperatures of up to 625°C (Fig. 12). But most of the pearlitic matrix had decomposed after annealing for 2 hours at 650° C (Fig. 13). Already for 625° C Brinell hardness had decreased from 189 to 171 kgf/mm², and for 650° C and 750° C it had dropped further to 132 and 121 kgf/mm², respectively.

Therefore, there are no objections to stress relief annealing if the temperature does not exceed 600° C.

Literatur/Reference
1) P. A. HELLER, H. JUNGBLUTH, Gießerei 42 (1955) 255/257

Metal Waves or Laking on Zinc-Based Diecastings

Gustav R. Perger and Peter M. Robinson

CSIRO, Division of Tribophysics
University of Melbourne
Parkville, Victoria, Australia

1. Introduction

Zinc-based diecastings for hardware and decorative applications require a surface finish which is virtually free from imperfections. The occurrence of some types of surface defects, such as cold shuts, can be avoided by adjusting operating parameters[1]. Other defects, however, occur persistently on certain types of casting; the surface defect termed metal wave, heat wave or laking is in this category. The lakes are areas encompassed by irregular lines or waves on flat or slightly contoured surfaces which are intended to look uniform. The position, shape and height of the waves which form the periphery of the laked area vary from casting to casting in a single production run. The waves are only visible by eye under certain angles of illumination; wave heights of 1 to 30 μm have been reported[2]. The laked areas have to be removed by polishing before the castings can be plated. This adds considerably to the overall cost of production.

Although "laking" is a common production problem, little information has been published on the nature of the defect and the conditions under which it occurs. It is considered generally that the defect arises from incorrect gating, insufficient control of operating conditions, or an incorrect balance between metal and die temperature, that is, almost any of the factors which control the quality of pressure diecastings. Wilcox[3] has proposed that metal waves can be eliminated by enlarging the area of the gate. On the other hand, Day[4] has suggested that the defect is most likely to occur when the design of the gate is such that the molten metal forms a skin across the die surface and then backfills during the re-

mainder of the injection cycle. This is in partial agreement with Bosley[5] who considered that metal waves are connected with preferential die filling and solidification. However, both Schneider[6] and Smith[7] found no relation between the gating design and the occurrence of metal waves.

The effect of machine operating parameters on laking have not been investigated in any great detail. Carrol[8] has recommended using as low an accumulator pressure as possible, consistent with overall casting finish, in order to reduce the occurrence of metal waves. Schneider[6] has reported similar success by reducing the accumulator pressure and this has been confirmed by Marchok and Draper[2]. The latter workers found that maladjustments of accumulator pressure, flow valve opening or oil level in the accumulator all contribute to the formation of metal waves.

Bosley[5] has suggested that metal waves are produced by high local die temperatures which lead to preferential solidification and to differential shrinkage of the surface away from the cavity walls. This is in agreement with the observations of Carrol[8] that the incorrect layout of water lines in the die is frequently responsible for metal waves on the castings. In addition, Smith[7] has reported the production of high quality castings after using a surface coolant spray and adding new water lines to dies which had yielded castings showing severe metal waves. However, in the two casting configurations tested, surface cooling alone gave little improvement. This is contrary to the observations of other workers[2][6][7][8] that spraying highly dilute lubricants on trouble-

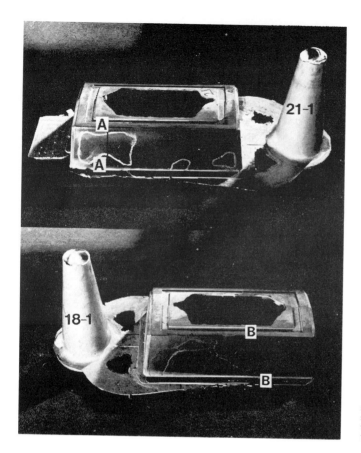

Fig. 1. Castings of name-plate holder showing the runner and overflow system, the outline of laked areas and the sections along which surface profiles were determined.

some areas tends to compensate for thermal imbalance which cannot be corrected otherwise.

It has been suggested also that the defect may be caused by the pulsating character of the injected metal stream coupled with the supercooling effect of the relatively cold cavity walls on the high velocity molten metal [2]. However, despite the conjecture as to the reasons for their occurrence, the physical characteristics of metal waves and lakes have not been defined in any detail [9] [2]. The object of the present paper is to investigate the metallographic characteristics of the laking defect in an attempt to define the factors which may control their formation.

2. Castings

The castings examined were of an automobile name-plate holder with two flat sides of approximately 113 cm² (Fig. 1). The cast-

ings were fed along both sides through gates with a total area of 170 mm². Production was on an E. M. B. 12 B machine, using a ram velocity during cavity fill of approximately 0.76 m/sec and a piston pressure of 0.5 MPa. The cavity fill time varied between 11 and 19 milliseconds. Because of limitation in the accuracy of the data collected, the coefficient of discharge into the cavity [1] [10] could not be calculated accurately; for most castings, however, it was between 0.3 and 0.6. The low coefficient of discharge at the gate was due partly to a restriction at the sprue which reduced the flow efficiency of the system and partly to the formation of blockages along the length of the gate (total gate length 240 mm).

During the casting trials, the dry shot speed [10] was varied over the range available on the machine and spray lubrication was used on the dies at irregular intervals. The metal temperature was maintained at 420 °C

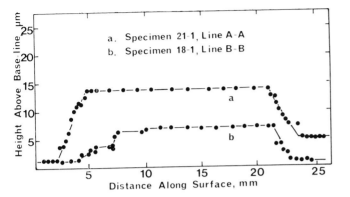

Fig. 2. Surface profiles through laked areas

and the bulk die temperature at 180 to 200 °C, this being normal industrial practice. All castings produced during the trial showed laking defects, the number and position varying from casting to casting.

3. Surface profiles

The flat surfaces of the castings were lightly polished and the surface profiles then measured using standard metrological techniques. The surface profiles along the lines A—A and B—B through laked areas on two castings (Fig. 1) are shown in Fig. 2. It is evident that the laked areas are plateaux, 5 to 15 μm above the general surface of the casting, bounded by gradual slopes. The latter rise typically 15 μm over 2.5 mm and appear visually as waves on the surface. The levels on either side of the plateaux were often different from one another. In extreme cases only one change in level was observed; the high area was usually adjacent to the gate or a side wall.

4. Metallography

One of the main problems in the metallographic examination of this type of defect is the small degree of surface relief and the comparatively long distance over which it occurs. This requires special techniques of illumination before any surface detail can be resolved. This has been overcome partly by the construction of a binocular microscope with a fluorescent illuminator ring of 16 cm diameter and a tilting and rotating stage. Figures 3 a and b show two views of the edge of a laked area on a lightly polished surface using this microscope. The interface is marked by fine surface porosity (Fig. 3 a)

which occurs on the lower slope of the plateau (Fig. 3 b). The difference in level between the laked area and the general surface of the casting and the occurrence of porosity along the interface was confirmed using the Nomarski interference contrast technique with a Zeiss microscope (Figs. 3 c and d).

The examination of metallographic sections through laked areas on the casting confirmed that the defect and the associated porosity was confined to the surface and did not extend into the bulk of the material. The resolution of any relevant microstructural detail close to the surface required the electroplating of samples with copper and iron for edge retention. A section through a typical laked area is shown in Fig. 4. The bulk of the casting consisted of evenly distributed primary dendrites of the zinc-rich phase in the eutectic mixture. At the surface, outside the laked area, large dendrites with a tendency towards columnar growth extended from the wall of the casting. These continued in an unbroken band under the laked area, the lake itself consisting of fine primary dendrites and the eutectic phase. It is evident from the microstructure that the lake area has solidified at a different time from the remainder of the casting surface.

5. Discussion

The experimental evidence indicates that the laked areas are actually plateaux above the general level of the casting surface, that there is porosity on the low side of the interface and that there is preferential soli-

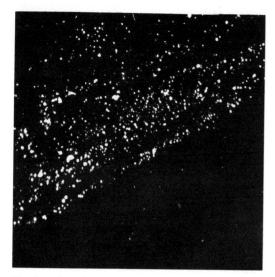

Fig. 3 a. Porosity of the low side of the interface, dark field illumination. 50 x

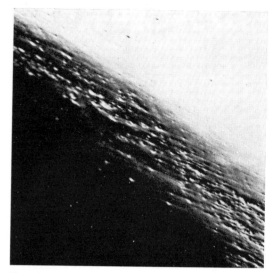

Fig. 3 b. Change in surface profile and porosity on the low side of the interface, oblique reflected illumination. 50 x

Fig. 3 c. Change in surface profile at edge of laked area, Nomarski Interference Contrast. 150 x

Fig. 3 d. Porosity at the interface, Nomarski Interference Contrast. 300 x

Figs. 3 a to d. The edge of the laked area on lightly polished surfaces

dification across the surface. The surface profile and the fine dendrite size within the lakes suggests that these areas solidified first under conditions of rapid heat extraction during initial filling of the die by metal streams. Subsequent filling and solidification of the surrounding surface results in the entrappement of gas to form fine porosity at the interface between the two areas. Initial solidification of areas outside the lakes occurs by the growth of large primary dendrites and these subsequently extend under the original solidified area of the lake. As the lake area is fed by molten metal during its solidification, the extent of shrinkage from the die cavity is less than that of the surrounding areas, resulting in the formation of a plateau.

The occurrence of lakes, therefore, appears

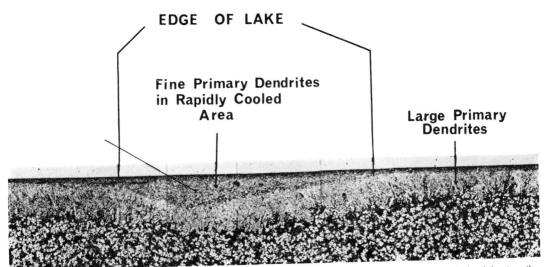

EDGE   OF   LAKE

**Fine Primary Dendrites
in Rapidly Cooled
Area**

**Large Primary
Dendrites**

Fig. 4. Section through a laked area showing the fine chill structure which forms the body of the lake together
with the large primary dendrites which extend on either side and below the lake. 160 x

to be due to preferential solidification across the casting surface, as suggested by B o s - l e y and others. Uneven die heating will obviously contribute to such an effect but consideration of the thermal balance in the die suggests that it should not be the prime cause in most production situations. It is more probable that, as is the present case, the occurrence of metal waves or laking is associated with a low efficiency of the total feeding system, as indicated by the discharge coefficient, and with long cavity fill times. Under these conditions only portions of the gate are operative and separate metal streams impinge on the cavity walls and solidify preferentially. Under conditions of slow cavity fill and high metal and die temperatures, preferential solidification results in the formation of lakes, while under opposite conditions the metal streams break up and form cold shuts.

It is considered, therefore, that the formation of metal waves and lakes depends primarily on the design of the gate and runner system and operating conditions. High flow efficiencies, with adequate feeding to all sections of the die, and short cavity fill times are desirable.

Acknowledgments

The castings used in this investigation were produced at Hucksons Diecastings, Springvale, Victoria, during a casting trial conducted by N. B u r y of Electrolytic Zinc Co. and A. D a v i s , CSIRO.

Summary

The nature of the casting defect known as metal waves or lakes which occur on the surface of zinc-based diecastings has been investigated. The lakes are actually plateaux above the general level of the casting surface, the waves which form the periphery of the lakes being the extended slope up to the plateau level. The laked area contains small primary dendrites and appears to be formed by the preferential surface solidification of the first metal streams to enter the die cavity. Subsequent filling and solidification of the surrounding surface results in the entrappement of gas to form fine surface porosity at the lower edge of the plateau and in the formation of a band of large dendrites extending beneath and on either side of the laked area.

It is suggested that the occurrence of lakes in the surface of castings is associated with low gate efficiency and long cavity fill times.

(Continued on the next page)

References

1) A. J. DAVIS, P. M. ROBINSON, The Production of Aluminium Diecastings Using a Fan Gate, Part 2. The Effect of Flow Conditions at the Gate on Casting Quality, Trans. Soc. of Diecasting Engineers (1975) Paper GT 75 122

2) R. P. MARCHOK, A. B. DRAPER, Elimination of Metal Waves in Hardware Finish Zinc Die Casting, Modern Castings 50 (1966) 66

3) R. L. WILCOX, Die Casting with Zinc Base Alloys, American Smelting and Refining Co., New Jersey (1961)

4) E. A. DAY, Other Factors and Problems to Consider in the Production of Good Quality Castings, Chap. LX in "A Visit with 500 Die Casting Plants", American Charcoal Co., Detroit (1959)

5) D. V. BOSLEY, private communication/persönliche Mitteilung

6) L. F. SCHNEIDER, Surface Defects in Die Casting, General Motors Third Year Report, General Motors Institute, Flint, Mich. (1958)

7) D. M. SMITH, Metal Waves in Zinc Die Casting, General Motors Institute, Flint, Mich. (1960)

8) R. J. CARROL, Some Factors Involved in Production of Hardware Finish Castings, Modern Castings 46 (1962) 101

9) S. D. SANDERS, W. D. KAISER, P. D. FROST, Metallographic Analysis of Zinc Diecastings, Trans. Soc. of Diecasting Engineers (1970) Paper 53

10) A. J. DAVIS, P. M. ROBINSON, The Production of Aluminium Diecastings Using a Fan Gate, Part 4. The Influence of Shot Velocity and Timing on Die Venting and Casting Porosity, to be published / demnächst

Cracks in Cylinder Blocks and in Cast Iron Cylinder Head

Friedrich Karl Naumann and Ferdinand Spies

Max-Planck-Institut für Eisenforschung

Düsseldorf

During the operation of tractors with cantilevered body, the lateral wall of the cylinder blocks cracked repeatedly. Three of the blocks were examined.

Figures 1 and 2 show one of the cases as seen from the outside and inside. The crack was always located approximately in the center of the sidewall and it ran in a horizontal direction. The cross section Fig. 3a showed more clearly that the thin case wall is reinforced at this location by a thicker rib. The crack was located in the thin-walled part next to the transition to the rib, as can be seen in larger scale in Fig. 4. At this location high residual stresses may be expected due to progressive solidification in the thin wall. In the plastic state during cooling of the casting the stresses evidently proved their effect already through a constriction of the wall cross section.

It remained to be examined whether the cracking was promoted by material defects. A chemical analysis of the castings gave the following composition:

The saturation point was calculated according to $\dfrac{C}{4.23 - 0.312 \cdot Si - 0.275 \cdot P}$. It was lower than 1 and proved that the cast iron was hypoeutectic. There are no objections to the composition.

Microsections of both cross sections were taken for metallographic examination. The grain structure of the thick-walled part consisted of uniformly distributed graphite of medium flake size in a basic mass of pearlite with little ferrite (Fig. 5). But the thin-walled part showed a structure of dendrites of precipitated primary solid solution grains with pearlitic-ferritic structure and a residual liquid phase with granular graphite in the ferritic matrix (Fig. 6). This structure contains a larger quantity of solid solution and therefore a higher concentration of carbon in the residual melt than would correspond to the equilibrium concentration. The structure is formed by undercooling of the residual melt. In this case it is promoted by fast cooling of the thin wall. It has a comparatively low strength.

C %	Graphite %	Si %	Mn %	P %	S %	S_C
3,30	3,01	2,48	0,66	0,104	0,054	0,96
3,29	not det.	2,44	not det.	0,100	not det.	0,96
3,30	not det.	2,28	not det.	0,168	not det.	0,95

Fig. 1. External view. 0.2 x

← crack site

Fig. 2. Internal view. 0.2 x

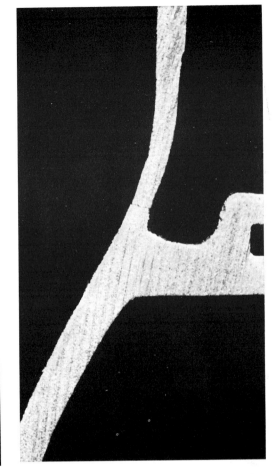

Fig. 3 a. Old design   Fig. 3 b. New design   Fig. 4. Section from Fig. 3 a. 1 x

Figs. 3 a and b. Sections through crankcase. 0.25 x

The fracture formation in the cylinder blocks can therefore be ascribed primarily to casting stresses. They could be alleviated by better filleting of the transition cross sections. The fracture was promoted by the formation of a undercooled microstructure of low strength in the thin-walled part. Therefore a reinforcement of the case wall is to be recommended. Shrinkage stresses as well as supercooling could be alleviated or avoided by prolonged cooling of the casting inside the mold.

Based on the recommendations given here, the design was changed (Fig. 3b) and cooling delayed. The microstructure of the rein-forced wall of a new case is shown in Fig. 7. It is clear that this is an improvement over the previous one (Fig. 6), but not as yet perfect. A further improvement could be achieved by inocculation of finely powdered ferrosilicon into the melt to facilitate nucleation and avoid supercooling.

Similar damage appeared in a cylinder head which had cracks in the water jacket. Figure 8 shows a section through this casting. The cracks emanated in part at the corners of the water chamber (i,i in Fig. 8), and in part at the outer surface (a,a). The first mentioned must be assumen to be stress cracks according to their location and they were

Fig. 5. Thick-walled part

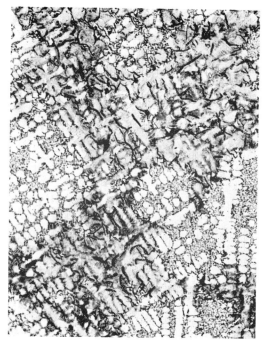

Fig. 6. Thin-walled part

Figs. 5 and 6. Case of old design

Fig. 7. Thin-walled part of reinforced case

Figs. 5 to 7. Microstructure of crankcase, etch: Picral. 100 x

Fig. 8. Sections through water chamber of cylinder head. 1 x

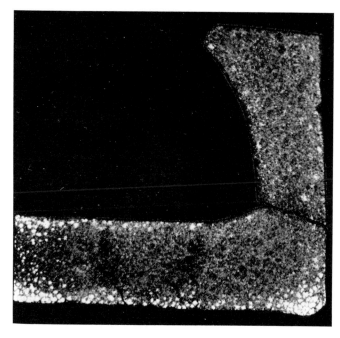

Fig. 9. Section of corner of cylinder head (lower right in Fig. 8). Etch: Picral. 3 x

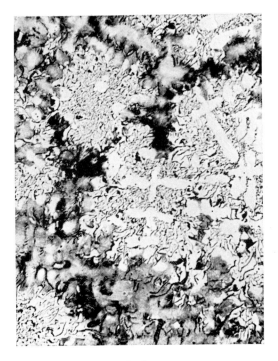

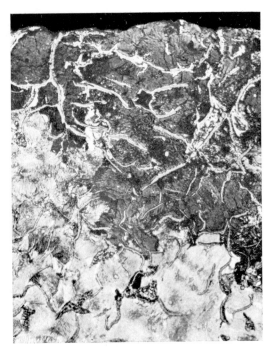

Fig. 10. Microstructure of edge

Fig. 11. Corrosion of edge

Figs. 10 and 11. Etch: Picral. 100 x

apparently already initiated during cooling of the casting. But this is improbable for the external cracks which did not penetrate the wall. Metallographic examination showed that they originated in ferrite clusters with granular graphite (Figs. 9 and 10). Their existence therefore was promoted by the supercooled structure. The question of their cause must remain open. In this connection the thought of operational vibratory stress may occur because it is well known that dynamic strength is decreased considerably in a ferritic bright border structure of low strength, such as is caused by surface decarburization. Incidentally, it may be mentioned here, even though it is of no consequence to the damage, that the inner wall was damaged by graphitic corrosion[1]) (Fig. 10). In order to avoid such damage, the same precautions should be observed as in the case described above.

Literatur/Reference

[1]) K. G. SCHMITT-THOMAS, G. FENZL, Der Maschinenschaden 38 (1965) 516
F. K. NAUMANN, F. SPIES, Prakt. Metallographie 4 (1967) 367/370

Fractured Swivel Head

Friedrich Karl Naumann and Ferdinand Spies

Max-Planck-Institut für Eisenforschung

Düsseldorf

The swivel head of a driving spindle of a four-high mill fractured. Plant inspection showed that the fracture originated in a darkly stained spot on the bottom of the cylindrical part and then continued into the cylinder walls in the two directions (Figs. 1 and 2). The fracture topography was of dendritic structure at the stained spot (Fig. 3). This led to the conclusion that a shrinkage cavity is present.

The section marked with chalk in Fig. 1 was cut from the bottom for examination and a cross section was made whose plane ran parallel to the fracture plane at a distance of 20 mm. In macroetching as well as in the Baumann print of this plane, phosphorous and sulfur segregations appeared as may be expected in the vicinity of a cavity (Figs. 4 and 5). The mount also showed microcavities which are characterized by their interdendritic position (Figs. 6 and 7).

Therefore the metallographic examination confirmed that the fracture of the swivel head was caused or favored by a cavity.

Fig. 1. Fractured swivel head

Fig. 2. Fracture origin (designated by arrow in Fig. 1). approx. 1 x

Fig. 3. Location from Fig. 2. 3 x

Fig. 4. Heyn-etch. 1 x

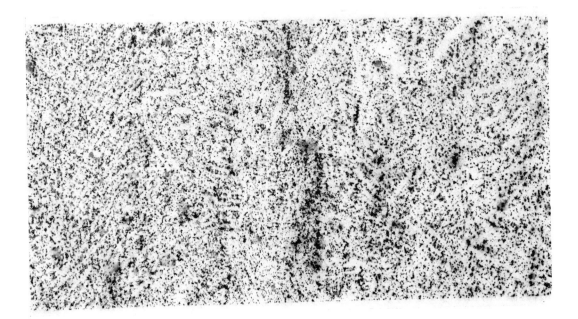

Fig. 5. Sulfur print according to Baumann

Figs. 4 and 5. Section parallel to fracture plane

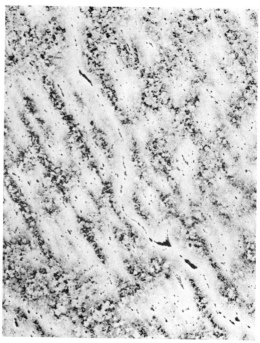

Fig. 6. Unetched. 50 x

Fig. 7. Oberhoffer etch (vertical illumination). 10 x

Figs. 6 and 7. Cavity in section according to Fig. 4

Damaged Impellers in a Rotary Pump

Egon Kauczor

Staatliches Materialprüfungsamt an der Fachhochschule
Hamburg

The two damaged impellers made of austenitic cast iron came from a rotary pump used for pumping brine mixed with drifting sand. On one of the impellers, pieces were broken out of the back wall in four places at the junction to the blades. The photograph in Fig. 1 shows that the fracture edges follow the shape of the blade. Numerous cavitation pits can be seen on the inner side of the front wall visible through the breaks in the back wall. The back wall of the as yet intact second impeller which did not show such deep cavitation pits was cracked in places along the line of the blades, as shown in Fig. 2.

The microstructure illustrated in Fig. 3 showing lamellar graphite and carbides in an austenitic matrix can be considered normal for the specified material GGL NiCuCr 15 6 2.

A specimen for metallographic examination was taken from one of the fracture edges. It was found that the material in this region was in the final stages of disintegration (Fig. 4). A further section was taken through an as yet unbroken region, the essential portion of which is shown at low magnification in Fig. 5. On the left marked ← cavitation pits can be seen on the internal surface

Fig. 1. Breaks in the back wall of impeller I and cavitation pits on the inner surface of the front wall. $^2/_3$ x

Fig. 2. Cracks in the back wall of impeller II following the shape of the blades. ²/₃ x

Fig. 3. Microstructure of a specimen from impeller I. Etched with V2A pickle. 500 x

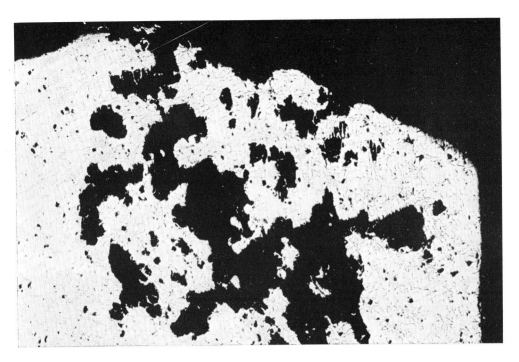

Fig. 4. Porous zone under a fracture edge of impeller I (shrinkage zone in Fig. 5). 20 x

Fig. 5. Cavitation pits (←) and porous zones (↘) in an unetched transverse section from an as yet unfractured region of impeller I. 2.5 x

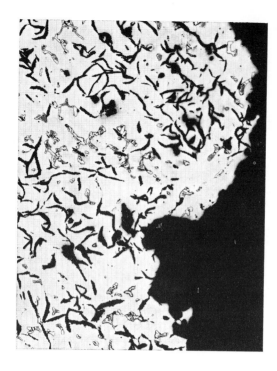

Fig. 6. Microsection from a cavitation zone. 100 x

of the front wall and on the right marked ↘ casting pores in the region corresponding to Fig. 4. It should be mentioned here that even when the specimes were cut carefully with a water cooled cutting wheel the stem in the brittle porous zone crumbled away and only at the third attempt was it possible to obtain a suitable whole specimen for examination. A section through a cracked region of the second impeller revealed a porous zone in the corresponding position.

Figure 6 shows a section of a cavitation region. The absence of corrosion products confirms that the surface here has been damaged by cavitation.

From the results of the investigation it can be concluded that the cause of the damage is the porosity at the junction between back wall and blades arising during the casting process. Cavitation has not contributed to fracture but could also have led to damage in the long term in the case of a sound casting. It is therefore advisable in the manufacture of new impellers to take care not only to avoid porosity but also to use alloy GGL NiCuCr 15 6 3 which has a higher chromium content and which is more resistant to cavitation.

Broken-Off Bearing Bosses of Scrap Shears

Friedrich Karl Naumann and Ferdinand Spies

Max-Planck-Institut für Eisenforschung

Düsseldorf

Three bearing bosses from the cover of scrap shears were sent in for examination. They had torn off the base plate to which they had been welded by fillet welds all around. Two of these were examined. They showed entirely the same symptoms. An

Fig. 1. View of broken-out bearing boss from below. ⅓ x

Fig. 2. Fracture view. 1 x

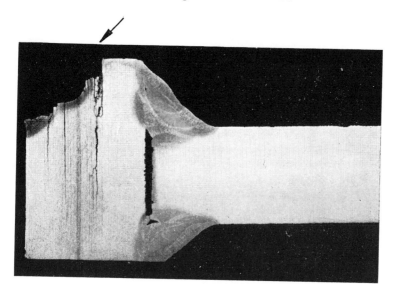

Fig. 3. Section through fracture (arrow), etch: Copper ammonium chloride solution according to Heyn. 1 x

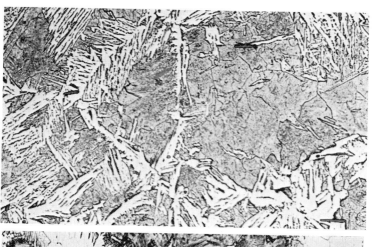

Fig. 4

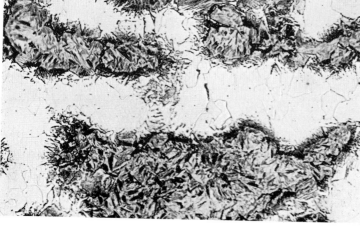

Fig. 5

Figs. 4 and 5. Structure of sheet next to weld seam, etch: Picral. 500 x

examination of the third one was meaningless because the fracture regions were so altered by the autogenous burning of the surfaces that were still adhering that only a very general material judgment could have been rendered. This could not have contributed anything to an explanation of the fracture cause.

Figure 1 shows one of the bosses seen from the underside. It had broken away on three sides along the welds. The cleaved fractures in the burned notches propagated partially above and partially below several incipient cracks which may be fatigue fractures (Fig. 2).

For metallographic examination sections were made across the fractures. From them it could be seen that the fractures had occurred either at the burned notches near the transition from the weld to the sheet, or else they run in the sheet material next to the weld (Fig. 3). The quality of the welds could not be judged because the opposite fracture pieces to which they adhered had not been sent in. The transition zones in the sheet material were very coarse-grained throughout (Fig. 4) and had partly transformed into martensite (Fig. 5). The hardness (HV 0.1) of these zones was 417 to 572 kp/mm² while it was about 130 kp/mm² for the sheet.

The break away of these bosses was at least favored by overheating and hardening.

Broken Structural Bolt

Friedrich Karl Naumann and Ferdinand Spies
Max-Planck-Institut für Eisenforschung
Düsseldorf

Two bolts from the stressed structure of a church building that had broken during stressing were examined to establish the cause of fracture. They had been ordered made of sigma steel 60–90, which is a heat treated low alloy structural steel with a yield point of at least 60 kp/mm² and a strength of 90 to 100 kp/mm².

The fracture of one of the bolts (1) showed that it had occurred in a double-Vee groove weld whose root was not completely welded (Fig. 1). The machining grooves at the end face of the head piece could still be seen in the center. Tinting of the fracture proved that the seam had cracked while exposed to heat. The other bolt (2) had cracked outside of the weld seam closely under the head. The fracture had a slightly fibrous structure (Fig. 2). Neither one had been particularly deformed before fracture. Figures 3 and 4 are lateral views of the bolts. The surface has been etched with copper ammonium chloride solution in order to bring out the weld seam. The weld rod material has not been attacked by the etchant, and therefore is highly alloyed. The unbroken weld seam of the second bolt was also not dense and had cracked. The welded-on head piece was shorter in this bolt than in the first.

Spectroscopic examination showed that the head pieces were made of manganese steel and the shafts of a chromium-molybdenum steel. Chemical analysis of bolt No. 2 showed the following values:

Thus the composition of the head piece corresponded approximately to the manganese steel 19 Mn 5 (Material No. 1 0845), a weldable construction steel with increased yield point and strength, while the shaft was made from a chromium-molybdenum steel of 42 CrMo 4 (Material No. 1 7225) according to DIN 17 200.

Figures 5 and 6 show the longitudinal sections through the weld seams, located in such a way that in one case the double-Vee seam had been cut through its entire width (Fig. 5) while in the other case only the root was cut across (Fig. 6). This also remained completely open in the second bolt. It can be seen already from the macroetching that the head pieces have been machined from the billets or plate in such a way that the fiber runs transversely to the direction of principal stress. Toughness is thus lowered considerably.

According to the fine microstructure the head pieces were untreated or normalized (Fig. 7), while the shafts were heat treated (Fig. 8). The previously mentioned unfavorable fiber direction of these pieces can also be seen from the secondary banded structure. The weld seams consisted of austenitic steel. They were permeated with hot tears (Fig. 9) as could be concluded already from the heat tinting of the fractures. The transition zones were hardened martensitically on both sides and had fractured in

	C %	Si %	Mn %	P %	S %	Cr %	Mo %
Head	0,19	0,50	1,50	0,026	0,034	0	0
Shaft	0,41	0,20	0,58	0,012	0,012	1,10	0,19

Fig. 1. Fracture of bolt 1 in double-Vee groove weld. 2 x

Fig. 2. Fracture of bolt 2. 2 x

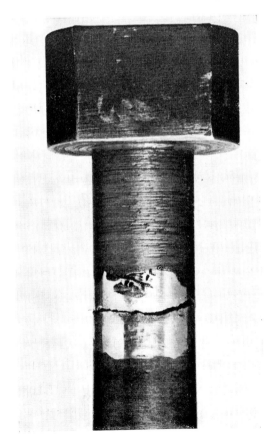

Fig. 3. Bolt No. 1

Fig. 4. Bolt No. 2

Figs. 3 and 4. Lateral views, surface etched according to Heyn. 1 x

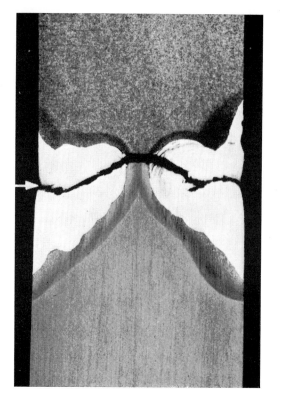

Fig. 5. Bolt No. 1

Fig. 6. Bolt No. 2, section plane turned 90°

Figs. 5 and 6. Longitudinal sections etched according to Heyn. 2 x

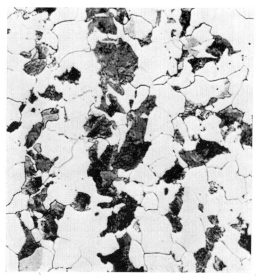

Fig. 7. Head pieces

Fig. 8. Shafts

Figs. 7 and 8. Microstructure of bolts, longitudinal sections, etch: Nital. 500 x

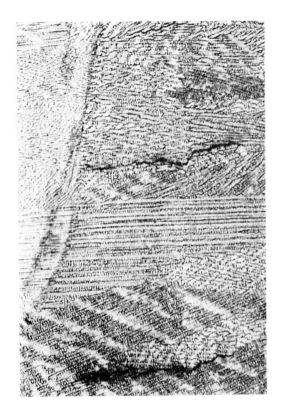

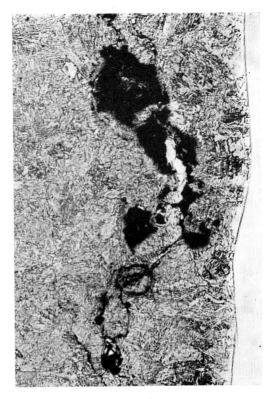

Fig. 9. Microstructure of weld seams, etch: V2A-pickling solution. 50 x

Fig. 10. Microstructure of transition zone with stress crack. Longitudinal section, etch: Nital. 100 x

places due to transformation stresses (Fig. 10).

The head pieces had a hardness of (HB 10/300/30) 179 to 186, or 167 to 168 kp/mm², respectively, corresponding to a strength of 62 or 57 kp/mm². Tensile test results with the heat treated shafts averaged 97 kp/mm² yield point, 109 kp/mm² tensile strength, 16 % elongation (δ_5) and 62 % reduction in area. These are normal values for the steel used.

The investigation showed that in this construction almost all errors that could be made had indeed been made, namely:

1. The bolts were not made from the alloy steel suitable for this purpose, but were welded together without planning of two steels unsuitable for this purpose, one of which had too low strength.

2. The austenitic weld seams showed hot tears and were not welded through to the root.

3. The pieces were not preheated before welding, a necessity in view of the strong hardenability of the steel, and they therefore had cooled too fast after welding, so that stress cracks had occurred in the transition zones.

This explains the fracture of bolt No. 1.

Bolt No. 2 very likely was overstressed during the impact caused by the breaking of the bolt No. 1. Its fracture was furthermore promoted by the unfavorable fiber direction due to wrong machining from the preformed material.

Fracture of a Cross on a Church Steeple

Friedrich Karl Naumann and Ferdinand Spies

Max-Planck-Institut für Eisenforschung
Düsseldorf

A cross crowned by a gilded cock on a church steeple hung in a slanted position from its support after a stormy night (Fig. 1). Figure 2 shows the construction of this support. Inspection showed that a fracture had occurred on the shaft of the cross which was formed by a seamless steel tubing of 60 mm O. D. and 2.7 mm wall thickness. The shaft was inserted into a supporting pipe of 12 mm wall thickness to a depth of about 3 m, and was firmly anchored in it. An additional support tubing of 650 mm length was inserted that fitted smoothly into the inner wall of the pipe and had welded-on annular bosses. This was for the purpose of reinforcement at the point of highest stress at the exit of the shaft from the supporting pipe. But the fracture had not occurred at this point of highest stress, but approximately 200 mm above it. A bell-shaped sheet metal cap was welded onto the shaft at this point. The tubing had fractured about 10 mm under this weld seam. The inserted support could not prevent the fracture but did prevent the toppling of the cross.

Figure 3 shows the fracture which was exposed after the bellshaped cap was opened up. The fracture occurred as rough jags around the tubing. The fracture plane was strongly abraded, so that neither the origin nor the character of the fracture could be established. A blueish discoloration could be seen in the fracture at the point closest to the weld seam. It was the size of a few square millimeters, and its position is marked by an arrow in Figs. 3 and 4. After the tubing was sectioned longitudinally it could be seen that it had been welded through at several points (Fig. 4).

An examination was conducted to find out whether a brittle fracture had occurred as a consequence of using an unsuitable steel. But an analysis showed that the steel of the shaft tubing contained only 0.033 % P and 0.004 % N, and thus could not be considered prone to brittle fracture or unsuitable for weld structures. Bending tests confirmed that the pipe had sufficient deformability. Nevertheless, it cannot be excluded that the tubing showed a brittle fracture under strongly multi-axial stresses. It is, however, more probable that a fatigue fracture had occurred.

For metallographic examination longitudinal sections were made through the planes designated as 6, 7, 8 and 9 in Fig. 4. They confirmed that welding caused complete melting of certain spots of the shaft tubing causing blistered or shrinkage cavities in the solidified area (Fig. 5 a). The blue spot in the fracture may possibly be due to such a cavity. It is shown in section 8–8 (Fig. 5 c). The investigation further disclosed that the shaft tubing had been welded in the fracture area fastening seam (Figs. 5 b and c). The reason for this remains unclear. It is noteworthy that the wall thickness of the tubing in this zone was reduced to 60 % of the original one. This occurred through grinding as can be seen from the grooves in Fig. 6. It may be assumed that traces of a previous weld seam were ground off in this spot. Such a weakening of the wall thickness alone could have been sufficient to cause the fracture of the shaft under dynamic loading that could be expected at this site and which may have been reinforced by the notch effect of the grinding grooves.

The investigation therefore showed that the design of the cross was an unfortunate mistake. If the bell-shaped cap was really essential it should have been fastened by other means than welding. Furthermore, the

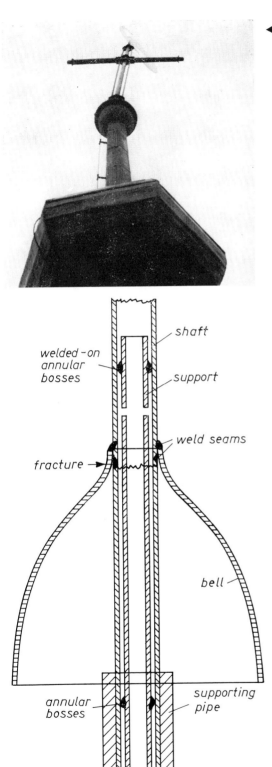

Fig. 2. Longitudinal section through support structures. approx. $^1/_5$ x

shaft

welded-on annular bosses

support

weld seams

fracture →

bell

annular bosses

supporting pipe

◀ Fig. 1. Damage site

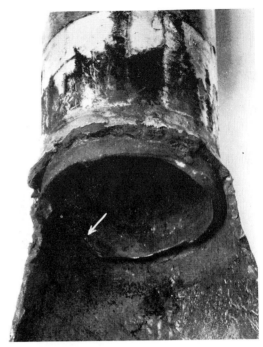

Fig. 3. Fracture. Blue-stained spot in fracture marked by arrow. Approx. $^2/_3$ x

6 7   8 9

6 7   8 9

Fig. 4. Interior view of fracture. Blue-stained fracture spot (left arrow), lower weld seam (right arrow). 1 x

welding was done poorly after an apparently initial aborted attempt. This then was the primary cause of fracture of the cross shaft.

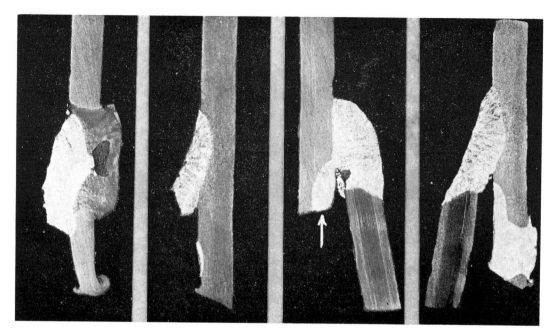

Figs. 5 a to d. Longitudinal sections in planes 6–6, 7–7, 8–8 and 9–9 according to Fig. 4, etch: according to Adler. Blue-stained fracture marked by arrow. 3 x

Fig. 6. Grind marks at exterior of the shaft tubing below weld fastening. 2 x

Cracking of Pipe Nipples in Welding

Friedrich Karl Naumann and Ferdinand Spies

Max-Planck-Institut für Eisenforschung

Düsseldorf

A number of seamless pipe nipples of 70 mm diameter and 3.5 mm wall thickness made of steel type 35.8 were oxy-acetylene welded to collectors of greater wall thickness with a round bead. X-ray examination showed crack initiation in the interior of the nipples close to the root of the weld seam. Welding was performed in a horizontal position. After tacking at two laterally facing locations, half the pipe circumference was welded from a starting point located at the bottom between the two tack-welds. The other half of the circumference was also welded from bottom to top in the opposite direction. Some overlapping of both half-beads at the beginning and the end provided for a sound juncture.

The cracks were always found in the proximity of the end crater, specifically on the side of the bead which was deposited first.

Figure 1 reproduces an X-ray film on which a crack can be detected which extends in both directions from the thickened end crater. Longitudinal specimens for metallographic examination were taken in the diametral planes A, B and C; of these plane B cuts into the terminal crater. Two additional metallographic sections radially displaced from B by 90° and 180° were prepared. Ground and etched sections A, B and C are reproduced in Fig. 2. The crack originates inside

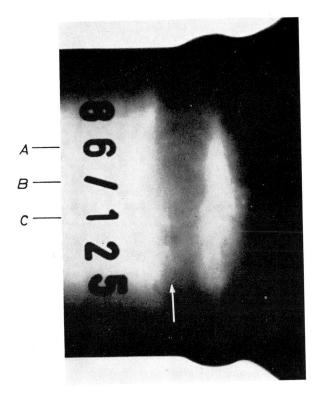

Fig. 1. Reproduction of an X-ray transmission film of a weld seam. A, B and C are section planes.

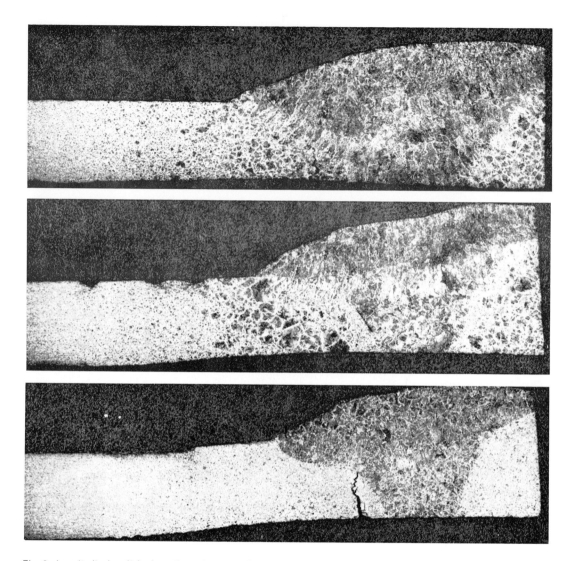

Fig. 2. Longitudinal polished sections A, B, C (from top to bottom) see Fig. 1. Etch: nital. Approx. 6 x

the base material close to the weld seam and terminates in its interface with the weld material. This takes place specifically where the V-seam widens appreciably as a result of depositing the second overlapping layer; in the crater itself the crack runs in the interface. It is noteworthy that a relatively fine-grained and ferrite-rich zone in which the cracks are situated lies immediately adjacent to the weld seam, and that this zone changes only at some distance from the seam into a coarse-grained region which is poorer in ferrite and exhibits a Widmannstaetten structure. Figures 3 a to c show the structure in these zones. A macro-etch after Oberhoffer

showed that the line structure of the pipe material ends with the beginning of the fine-grained crack region (Fig. 4). Smaller initial cracks were also found at the origin of the weld seam; the intermediate specimens were however free of cracks. The cracks therefore only appeared where the originally deposited bead was remelted in the regions of overlap.

The cracks apparently occur at the austenite grain boundaries (Fig. 5). Their edges are oxidized (Fig. 6). We are thus dealing with typical thermal cracking. The fine precipitates in the ferritic zones around the cracks (Fig. 7)

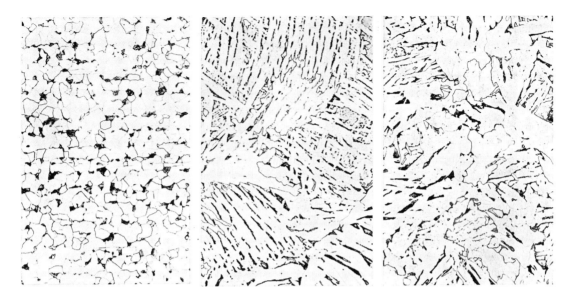

Fig. 3 a. Unaffected pipe material   Fig. 3 b. Coarse-grained region   Fig. 3 c. Crack region

Figs. 3 a to c. Structure in polished section B; see Figs. 1 and 2, Etch: nital. 200 x

Fig. 4. Polished section C; see Figs. 1 and 2, etch after Oberhoffer. 10 x

are vitreous silicates, based on their behavior in polarized light. This form of selective oxidation is favored by the scarcity of oxygen in narrow cracks. In the extensions of the cracks, ferritic bands are often recognizable (see Fig. 5); their origin is probably due to the inclusions precipitated at the austenite grain boundaries (Fig. 8). These inclusions are optically isotropic and, judging by their shape and color, appear to be sulfides of manganese. These manganese sulfides exhibit an elongated shape in the longitudinal section taken from the pipe material which has not been thermally affected by the welding (Fig. 9 a). In the proximity of the weld seam they have transformed into shorter particles while still preserving their arrangement in lines (Fig. 9 b). Immediately adjacent to the weld seam the manganese sulfides have evidently gone into solution completely

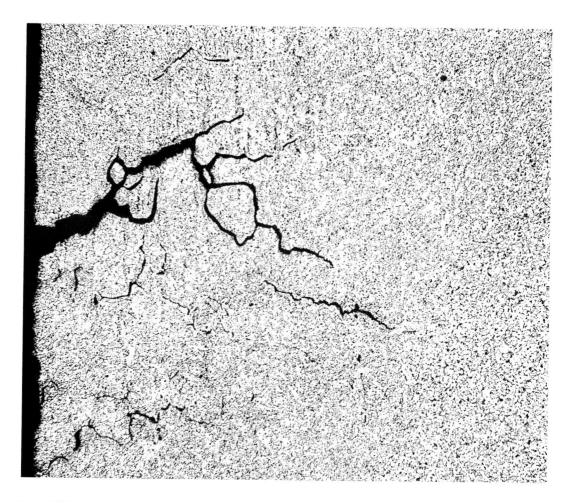

Fig. 5. Weld crack in a section through the end crater, Etch: nital. 50 x

and during cooling have precipitated at the boundaries of the coarse grains of austenite (Fig. 9 c). The distinct grain size in the proximity of the weld seam (see Figs. 3 a to c), may be related to this differing distribution of the inclusions.

The origin of the cracks may therefore be visualized approximately as follows: During the deposition of the first bead a weakening of the grain boundaries of the coarse austenite or delta-ferrite grains took place in the highly overheated regions next to the weld bead. This weakening was due to the precipitation of manganese sulfides or manganese oxysulfides; possibly with the participation of picked-up oxygen. Upon welding the opposite side, the grain boundaries thus damaged, cracked apart under the effect of the thermal stresses. E. H o u d r e m o n t[1] has designated this condition as overheating with grain boundary damage; it may be removed by heat treatment if it has not already caused crack formation as in the present case.

It has been shown that weld cracking is a form of hot shortness. It appears at temperatures between 1000 and 1350° C, preferably over 1200° C [2]. The occurrence of weld cracking is favored by higher contents of carbon as well as phosphorus, hydrogen, oxygen and particularly sulfur [3]. A test revealed that the sulfur content of the steel used was 0.045 to 0.050 % and lay at the

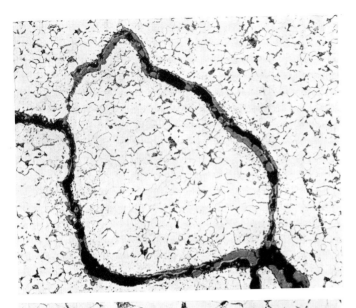

Fig. 6. As Fig. 5. 200 x

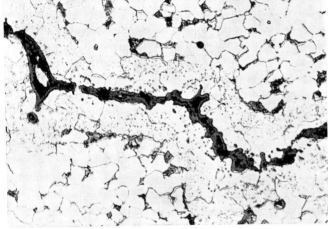

Fig. 7. As Fig. 5. 500 x

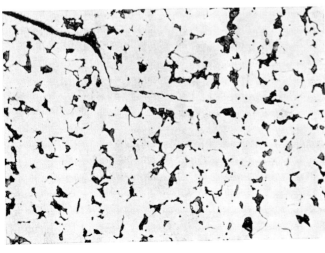

Fig. 8. MnS inclusions and ferrite network at the austenite grain boundaries in longitudinal section through the starting point of the weld seam, Etch: picral. 500 x

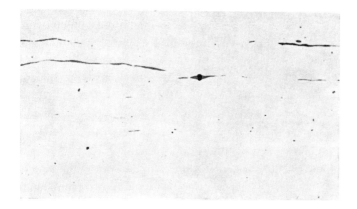

Fig. 9 a. In the unaffected part

Fig. 9 b. In the proximity of the weld seam

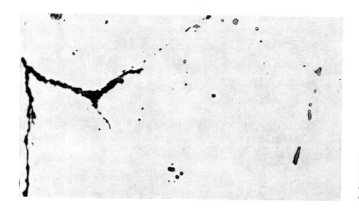

Fig. 9 c. In the crack region

Figs. 9 a to c. Change of shape and distribution of the manganese sulfide inclusions as the weld seam is approached, longitudinal polished section, unetched. 500 x

upper boundary of the admissible range. Given the construction and welding technique used, it would certainly have been preferable to make the nipples of a steel lower in sulfur content. However, by taking advantage of all the potential in shaping and welding technology, it should be possible to prevent crack formation with steel type 35.8 of normal composition.

[1] E. HOUDREMONT, Stahl u. Eisen 72 (1952) 1536/40
[2] A. ANTONIOLI, Stahl u. Eisen 62 (1942) 540/45
[3] K. L. ZEYEN, W. LOHMANN, Schweißen der Eisenwerkstoffe, 2. Aufl. Düsseldorf (1948)

Welded Pipes with Hard Spots

Friedrich Karl Naumann and Ferdinand Spies

Max-Planck-Institut für Eisenforschung

Düsseldorf

The pipes were made of low-carbon Thomas steel which had been welded longitudinally employing the carbon-arc process with bare electrode wire made for argon-shielded arc welding. Difficulties were encountered during the cutting of threads because of the presence of hard spots. The seams showed crater-like indentations on the hard spots (Fig. 1) which left shiny worn areas during the cutting of threads (Fig. 2). Their hetero-geneous structure is shown macroscopically in Fig. 3 which represents a longitudinal section through the seam. As can be seen from Figs. 4 to 8, the structure of the welding seam is composed of a number of different microstructures of different carbon content and at different transformation stages.

Figure 4 represents the normal microstructure of the weldseam which is the overheated

Fig. 1. Mid-portions

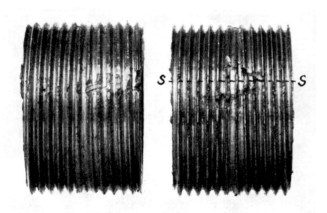

Fig. 2. Thread-portions

Figs. 1 and 2. Views of the tube specimens. 1 x

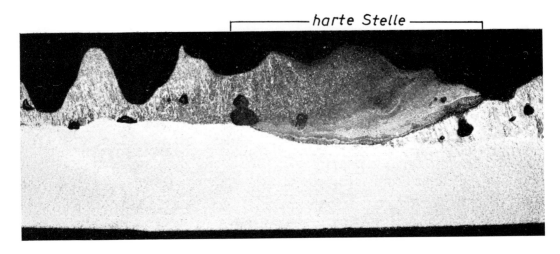

Fig. 3. Longitudinal section s---s in Fig. 2 through a hard spot. Etchant: Nital. 10 x

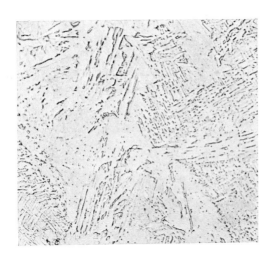

Fig. 4. Normal location

Figs. 4 to 8. Microstructure in the weldseam, longitudinal section. Etchant: Nital. 500 x

structure of a steel low in carbon. The area in Fig. 5 is already carburized considerably but the steel is still hypo-eutectoid as indicated by the precipitation of ferrite at the austenitic grainboundaries. Further transformation has taken place in the pearlite stage and partly into the intermediate and martensitic stages. The area shown in Fig. 6 is free of ferrite, thus approximately eutectoid; the microstructure consists of pearlite and martensite with remnants of austenite. Figure 7 reproduces the microstructure of a location with hyper-eutectoid carbon content. It consists of secondary cementite in the shape of long needles and grainboundary precipitates and pearlite. In view of the presence of cementite nuclei, an undercooling of the $\gamma-\alpha$ transformation into the intermediate or martensite stage did not take place. In some areas, as shown by Fig. 8, carburization has extended into the region of hypo-eutectic cast iron as proven by the presence of ledeburite. Here also the basic microstructure consists of pearlite. Accordingly, the weldseam has carburized to over 2 % C in places with consequent hardening. Even though it is of no significance for the present problem, let it be mentioned

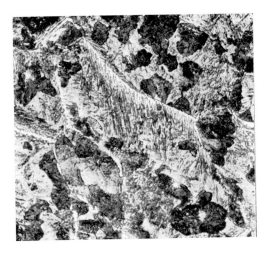

Fig. 5. Hypo-eutectoid

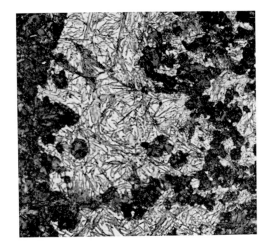

Fig. 6. Near-eutectoid

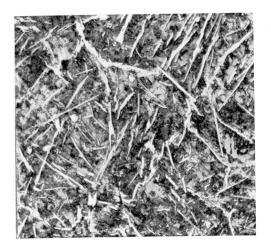

Fig. 7. Hyper-eutectoid

Figs. 5 to 8. Carburized locations

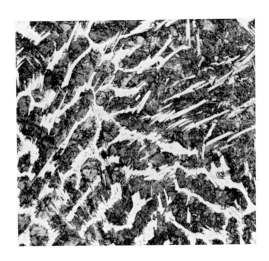

Fig. 8. Hypo-eutectic cast iron

in passing that the weldseam is permeated with gas bubbles.

Accordingly, the welding conditions were such that a carburizing atmosphere could develop which led to an increase in carbon content and hardening at certain locations such as terminal bells and lap joints. This explains the processing difficulties during the threading operation.

Investigation of Superheated Steam Push Rod Spindles

Friedrich Karl Naumann and Ferdinand Spies

Max-Planck-Institut für Eisenforschung

Düsseldorf

A spindle made of hardenable 13 % chromium steel X 40 Cr 13 (Material No. 1.4034) that was fastened to a superheated steam push rod made of high temperature structural steel 13 CrMo 44 (Material No. 1.7335) by means of a convex fillet weld, fractured at the first operation of the rod directly next to the weld bead. The carbon-rich stainless spindle steel was said to have been selected because of its good operating properties.

At the fracture a narrow circumferential zone of fine fracture grain could be observed that pointed to a hardening of the material next to the seam.

A longitudinal section through the fracture origin after various etches showed the structure presented in Figs. 1 and 2. A low alloy additive was used for welding that was alloyed in contact with the spindle to such an extent that it was attacked by copper ammonium chloride solution according to Heyn, but not by Nital. In this alloyed area of the weld seam, as well as in the spindle directly next to the seam, several short cracks appeared which are designated by arrows in Fig. 1. Such a crack may also have been the cause of fracture. Figure 3 shows the crack in the transition zone of the spindle material on a larger scale. This zone has a hardened microstructure of martensite and ferrite (Fig. 4), while Fig. 5 shows the contrasting annealed microstructure of the unaffected spindle material consisting of chromium carbides in a ferritic matrix.

Based upon the results of this investigation, the works manager of the power station in which several more of these shutoff valves were used, sent a push rod that did not show any such damage as yet, for non-destructive testing for cracks. It is shown in Fig. 6. The spindle is screwed into the foot of the pusher and secured against torsion by two bolts. It is not clear what sense the welding has made under these circumstances. In any case, it certainly was unsuitable as a means for rounding off the cross sectional transition as was proved by the fracture caused by it.

Figure 7 shows the weld seam to be examined in actual size. Without any special aids it can easily be recognized that the seam has fractured to a length of approximately 30 mm in a circumferential direction. No further damages could be detected even by magnetic powder testing. Based upon this finding, the spindle was opened lengthwise through the fracture location. As can be seen from Fig. 9 the crack propagation pointed to the fact that the crack originated at the inside. A short incipient crack of 0.2 mm depth could also be found in the spindle material next to the weld at location a. Incipient cracks in the weld seam and adjoining it in the spindle material were also found on the side opposite the crack (Fig. 8, locations a and b). They are shown on a larger scale in Figs. 10 and 11. Similar, but shorter incipient cracks were also found in the weld joint with which the collar that secures the seal when the pusher is open is fastened onto the spindle (compare Fig. 6). The structure in the cracked zones consisted of martensite with little ferrite, just like in the broken spindle.

This investigation has shown that the fracture of the superheated steam push rod spindle was caused by hardening and

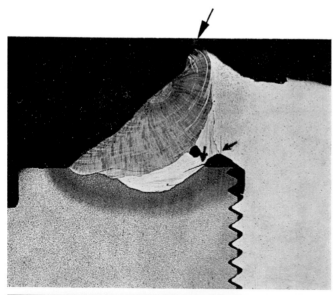

Fig. 1. Etch: Nital

Fig. 2. Etch: Copper ammonium chloride solution according to Heyn

Figs. 1 and 2. Longitudinal section through weld seam of fractured spindle. 3 x

hardening crack formation in the weld seams and adjoining areas. Several errors were made during design and production of the push rods.

1. It would have been preferable to avoid welding near the cross sectional transitions altogether in consideration of the crack sensitivity of high hardenability steels.

2. If for some reason this was not possible, then all precautions should have been taken that are applicable to the particular steel, such as preheating, slow cooling and stress-relief tempering after welding.

3. The selection of an austenitic additive material should have been considered because it could have equalized stresses due to its high elongation.

4. Most probably, however, a material of lower hardenability should have been selected for the spindle if high operating properties were of paramount importance.

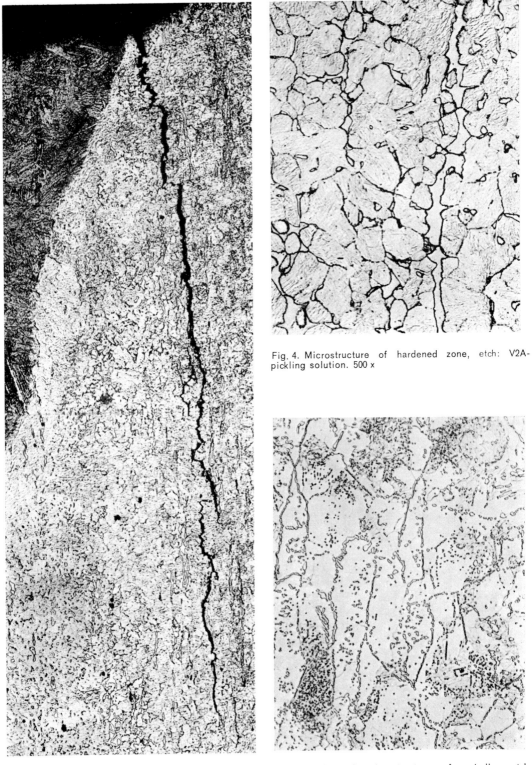

Fig. 3. Stress crack in hardened zone of broken spindle (left: weld), longitudinal section, etch: V2A-pickling solution. 100 x

Fig. 4. Microstructure of hardened zone, etch: V2A-pickling solution. 500 x

Fig. 5. Unchanged microstructure of spindle, etch: V2A-pickling solution. 500 x

Fig. 6. Superheated steam push rod with unbroken spindle. ¼ x

Fig. 7. Section of Fig. 6, weld seam with crack. 1 x

Fig. 8. Side opposite crack

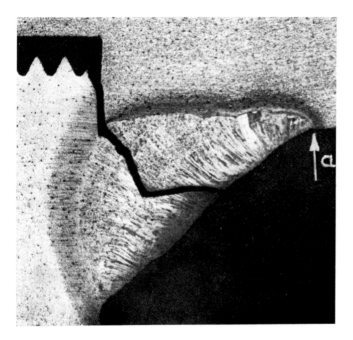

Fig. 9. Section through crack

Figs. 8 and 9. Longitudinal section through weld seam of unbroken super-heated steam push rod, etch: V2A-pickling solution. 5 x

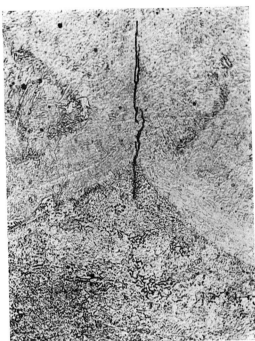

Fig. 10. Crack in transition from weld seam (left) to spindle (location a in Fig. 8)

Fig. 11. Crack in weld seam (location b in Fig. 8)

Figs. 10. and 11. Longitudinal section, etch: V2A-pickling solution. 100 x

Investigation of Worn Chain Links

Friedrich Karl Naumann and Ferdinand Spies

Max-Planck-Institut für Eisenforschung

Düsseldorf

Three links of a chain showing unusually strong wear were examined to find the damage cause. Corresponding to the stress, the wear was strongest in the bends of the links, but it was especially pronounced in the bend in which the butt weld seam was located (Fig. 1).

Transverse and longitudinal sections were made for metallographic examination from the straight undamaged parts of the links. The investigation showed that the links were manufactured from an unkilled carbondeficient steel and they were case hardened to a depth of 0.8 to 0.9 mm (Figs. 2 and 3).

The peripheral structure at the places not showing wear consisted of coarse acicular martensite with a high percentage of retained austenite (Fig. 4). The links were therefore strongly overheated, probably directly heated during case hardening. The butt weld seames were not tight and were covered with oxide inclusions (Fig. 2).

From the fact that wear occurred preferentially at the welds it may be concluded that this weld defect contributed to the substantial wear. Probably the carburizing medium penetrated into the partially open joint gap and carburized the overlapping tongue. During

Fig. 1. Chain link with strong wear at welded end. 1 x

Fig. 2. Longitudinal section through weld, etch: copper ammonium chloride solution according to Heyn. 1 x

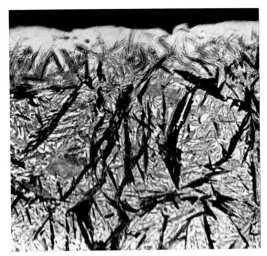

Fig. 4. 500 x

Fig. 3. 100 x

Figs. 3 and 4. Peripheral structure of a site free of wear, longitudinal section, etch: Picral

deformation of the core material splinters then broke out of the high temperature embrittled case layer through overloading and through impact blows that occur during stressing of chains during dissolving of slings. These splinters had a grinding effect on the adjacent links at the points of contact. The extremely strong wear could thus be explained.

This leaves unanswered the question whether these chains could have withstood the high operating stress if they had been welded satisfactorily and hardened correctly, and whether it makes any sense at all to case harden highly stressed chains of this type.

Fractured Post of a Loading Gear

Friedrich Karl Naumann and Ferdinand Spies

Max-Planck-Institut für Eisenforschung
Düsseldorf

In a shipyard one of the two posts of a loading gear fractured under a comparatively small load at the point where it is welded into the ship's deck. The post consisted of several pipe lengths that were produced by longitudinal seam welding of 27 mm thick sheets. The sheet metal was a construction steel of 60 to 75 kp/mm² strength.

The selection of the pipe length to be examined contained the fractured circumferential seam. The fracture was so well hidden that its course was made visible only after etching with dilute hydrochloric acid (Fig. 1). It corresponded almost exactly to the transition of the welding seam into the pipe.

A cross section through the welding seam was taken for metallographic examination. The etched section for macroscopic observation is shown in Fig. 2. From it can be seen that the crack in the zone under the seam affected by the welding heat is much more pronounced than at the surface. It probably originated in this zone.

Microscopic examination showed in part a

Fig. 1. Surface of cracked pipe length (arrow) removed from ship's deck. 1 x
(Above and below: Pipe mast, center: Remaining ship's deck with burn traces)

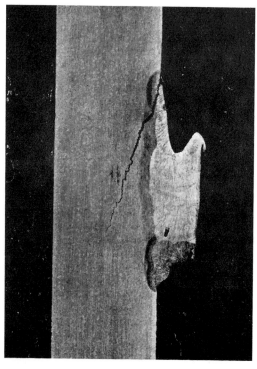

Fig. 2. Longitudinal section across welding seam. Etch: Copper ammonium chloride solution according to Heyn. 1 x

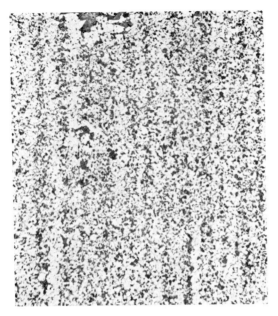

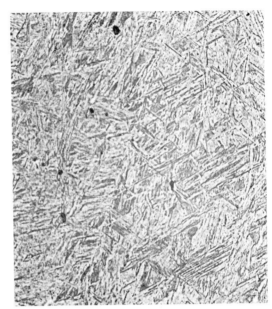

Fig. 3. View of part not subject to welding heat. 100 x

Fig. 4. Transition to welding seam. 500 x

Figs 3 and 4. Structure of pipe material. Longitudinal section, etch: Picral

purely martensitic structure (Fig. 4) in the transition zone of the pipe material, that originally had a fine-grained structure of ferrite and pearlite (Fig. 3). Hardness (HV 5) in this zone reached values from 376 to 401 kp/mm² when compared to 199 to 206 kp/mm² in the original pipe material. Therefore crack formation occurred due to transformation stresses as a consequence of fast cooling after welding.

The fine grained raw material had an elongation of 21.0 % (δ_{10}) at a yield point of 42.7 kp/mm² and 66.7 kp/mm² tensile strength. It had 63 % reduction of area and 16.1 and 15.2 kpm/cm² notch toughness at 18 and 0 °C, respectively. Therefore it corresponded to expectations in all respects.

Thick-walled parts of steels of such high strength must be preheated to approximately 200 °C along the edges prior to welding in order to minimize the strong heat losses by the cold mass of the part. In the case under investigation this either was not done at all or the preheating was not high enough or sufficiently uniform.

This damage was therefore caused by a welding defect.

142

Leaky Socket Pipe from the Safety Return Circuit of a Heating Installation

Egon Kauczor
Staatliches Materialprüfungsamt an der Fachhochschule
Hamburg

The elbow made from welded steel tube had become leaky along a well defined line in the axial direction. In order to examine the internal wall of the tube, the elbow was cut open in the longitudinal direction. Figure 1 shows one half of the elbow with the split. It can be seen that the entire wall of the tube is corroded and that the longitudinal welded seam stands out clearly as a result of particularly intensive corrosive attack. The appearance of the corroded surface indicates the action of water with a high oxygen content. The oxygen in the return water must originate from the ventilation of the open expansion vessel.

In order to explain the preferential corrosive attack in the zone of the longitudinal welded seam, a cross section was taken at a point at which the weld zone looked the soundest. The etched microsection in Fig. 2 shows that the butt weld had been defective from the beginning. A bonding defect filled with oxide runs from the outside deep into the wall of the tube. A similar welding defect had presumably been present in the region of the wedge shaped corrosion furrow on the internal tube wall. Due to the corrosion favouring effect of this crevice, water with a high oxygen-content that was perhaps still warm or even hot found particularly favourable conditions for corrosion in the defective welded seam (crevice corrosion). The tube material itself is perfectly satisfactory and in no way responsible for the failure.

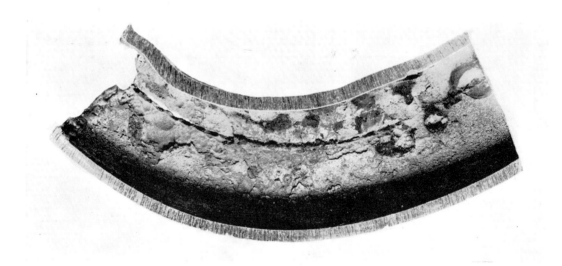

Fig. 1. Heavily corroded internal wall and split in one half of a longitudinally sectioned elbow. 1 x

Fig. 2. Cross section through the welded seam etched with 2 % nital. 50 x

Cracks in Flame Hardened Operation Handles

Egon Kauczor

Werkstoffprüfamt der Freien und Hansestadt
Hamburg

Operation handles produced from C45 steel showed many fine cracks at the flame hardened noses. In Fig. 1 the hardened zone of one of the samples has been made visible by an etch using 10 % nital. The cracks run from the corners of the marked indentations caused by the tool during alignment as shown in Fig. 2 at a larger magnification.

The Vickers hardness was determined at a load of 10 kgf. Values HV10 of 630 kgf/mm² and 214 kgf/mm² were obtained at the nose and the original material, respectively. The hardness of the nose reaches the upper limit of the hardness obtainable by hardening C45 while the hardness of the material in the ori-ginal state corresponds to a normalized structure.

The results of the metallographic investigation are shown in Figs. 3 and 4. The ferritic-pearlitic structure of the parts not influenced by flame hardening is fine grained corresponding to the normalized state (Fig. 3), while the martensitic structure of the nose shown in Fig. 4 is very coarse as compared to a normally hardened structure. This points out that the nose has been overheated during flame hardening. The orientation of the coarse martensite needles indicates the size of the austenite grains before quenching.

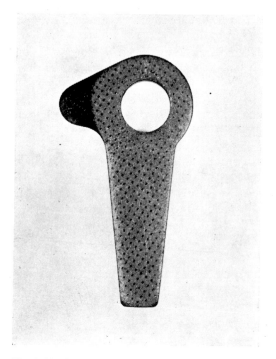

Fig. 1. Handle etched with 10 % nital to show the flame hardened zone (dark). 1 x

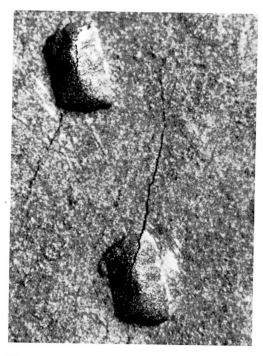

Fig. 2. Cracks running across the indentations caused by the tool during alignment. (Photographed with the TESSOVAR, Carl Zeiss, Oberkochen). 37 x

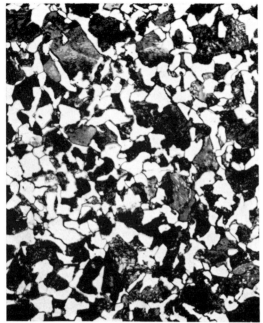

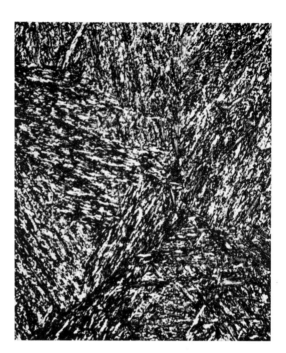

Fig. 3. Structure of the original material, etched with 2 % nital. 500 x

Fig. 4. Structure of the flame hardened top. 500 x

From these investigations it can be concluded that the numerous hardening cracks have been caused by abrupt quenching from over-heating temperature and by local stress concentrations due to indentations of the tool caused during alignment.

Screen Bars Destroyed by Intergranular Corrosion

Friedrich Karl Naumann and Ferdinand Spies
Max-Planck-Institut für Eisenforschung
Düsseldorf

Fragments of screen bars which as structural elements of a condenser had come into contact with cooling water from the mouth of a river were received [1]. Reportedly they were made of the stainless austenitic chromium-nickel-molybdenum steel X 5 CrNiMo 18 10 (Material No. 1.4401). According to an analysis by the Association for Technical Surveillance the cooling water at a pH-value of 6.6 to 6.7 contained 10.000 mg Cl' and 1440 to 1520 mg SO_4'' per liter. An investigation was to determine whether the bars consisted of the specified alloy; it was also to reveal the type and origin of the damage. Enquiries were also made if steel of type X 5 CrNiMo 18 10 was suitable for this application or what substitute material should be employed in the repair work.

The bars were fractured repeatedly (Fig. 1). In contrast to a former investigation [1], the ruptures did not occur exclusively or even preferentially at the loops of the bars, but just as frequently at locations between them.

On some of the bars, welding globules had again deposited. A number of bars showed cracks in addition to the fractures (Fig. 2) and individual sites of corrosion pitting (Fig. 3). The fracture surfaces showed a grainy structure (Fig. 4) in contrast to those formerly investigated [1]. This indicates that fracture was not preceded by appreciable deformation.

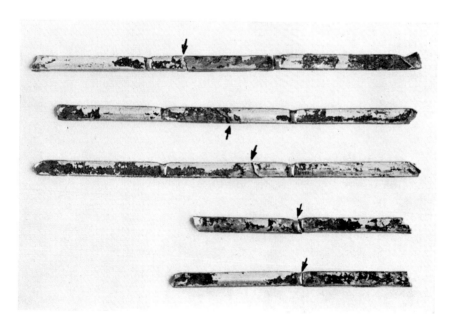

Fig. 1. Fractured screen bars. 1 x

Fig. 2. Bar with cracks. 15 x

Fig. 3. Bar with corrosion pitting. 15 x

Fig. 4. Fracture. 15 x

Analysis of the bars yielded the following mean values [wt.-%]:

C	Cr	Mo	Ni	Ti
0,077	17,85	2,31	12,73	0,05

With the exception of a slight excess in carbon content, this composition corresponds to the specification for X 5 CrNiMo 18 10.

Longitudinal and transverse sections through the defective regions revealed that the fissures are the result of intergranular corrosion (Figs. 5 to 7). This occurs in austentitic steels when they have been annealed at temperatures from 600 to 750° C and carbides have segregated at the grain boundaries; a process favored by cold work prior to annealing. The presence of such precipitates is indicated in Fig. 8 by the strong shading of the grain boundaries. At higher magnification (Fig. 9) the precipitates themselves become visible. The fact that they are attacked by electrolytic etching in ammonia water (Fig. 10) proves that carbides of high chro-

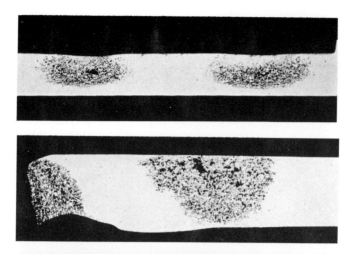

Fig. 5. Longitudinal sections through defective regions, unetched. 5 x

Fig. 6. Transverse sections through defective regions, unetched. 5 x

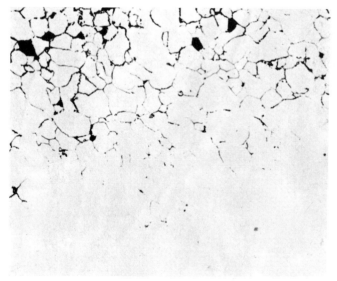

Fig. 7. Transverse section through the transition zone between the disintegrated region and the corrosion-free region, unetched. 100 x

mium content or chromium carbides are present. The regions near the grain-boundaries which are depleted in chromium as a result of this carbide precipitation can be attacked by corrosion or dissolved out (Fig. 11), which causes the alloy to lose its cohesion. This type of corrosion is therefore known as grain disintegration or intergranular corrosion. The pronounced localized concentration of the attack on individual regions is noteworthy; it is characteristic in the corrosion of alloys whose stability depends upon the formation of a passive layer. The carbides can be redissolved at a temperature of

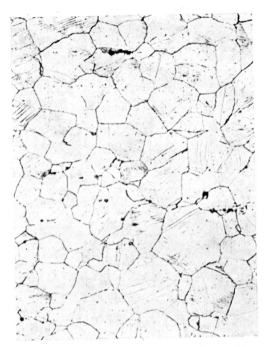

Fig. 8. Etching treatment: V2A-etching solution. 200 x

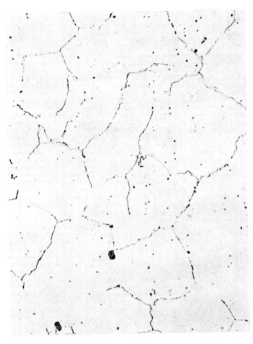

Fig. 9. Etching treatment: 50 % aqueous solution of nitric acid, 2 V 3 min. 500 x

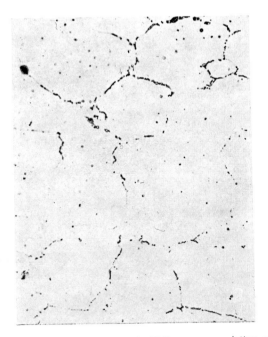

Fig. 10. Etching treatment: 10 % aqueous solution of ammonia, 2 V 3 min. 500 x

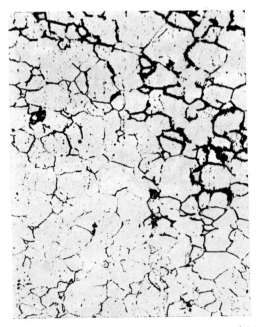

Fig. 11. Structure of transition zone to region of intergranular corrosion. Etching treatment: 50 % aqueous solution of nitric acid, 2 V 3 min. 100 x

Figs. 8 to 10. Structure of corrosion-free region

Fig. 12. Structure after heat treatment (½ hr. at 1050° C/ water). Etching treatment: V2A-etching solution. 200 x

1050° C and maintained in solution by a rapid quench; this will cause the steel to lose its tendency for intergranular corrosion (Fig. 12).

Therefore the fault that was made in this case consisted in annealing the steel at a temperature in the critical region. This was probably done to relieve the stresses which had originated during the cold-forming of the bars and had led to the damage by stress-corrosion described earlier [1]. This would have been the correct method for a ferritic steel. However this austenitic steel requires the special heat treatment indicated. When an anneal in the critical region is unavoidable and the indicated additional treatment is impossible or difficult, a type of steel has to be chosen which is resistant to intergranular corrosion, i. e. an alloy in which the carbon is bound to e. g. titanium or niobium as a carbide of low solubility. The following steels should be taken into consideration for such an application: X 10 CrNiMoTi 18 10 (Material No. 1.4571) or X 10 CrNiMoNb 18 10 (Material No. 1.4180).

[1] F. K. NAUMANN, F. SPIES, Prakt. Metallographie 9 (1972) 592/596

Investigation of a Case Hardened Sleeve

Friedrich Karl Naumann and Ferdinand Spies
Max-Planck-Institut für Eisenforschung
Düsseldorf

A heat treating shop sent a case hardened sleeve made of C 15 (Material No. 1.0401) for determining the cause of fracture. The sleeve was flattened at two opposing sides and had cracked open at these places, the crack initiating at a face plane. The wall of the sleeve was 9 mm thick, but the flat ends were machined down to 5.5 mm from the outside. The customer had specified a 2 mm case depth and a hardness of at least 55 RC at a depth of 1.5 mm.

The etched cross section in Fig. 1 of the cracked end shows that a case layer had a depth of 2.3 mm so that the sleeve was almost through-hardened at the flat ends. While the core material with the full wall thickness had the quench structure of low carbon steel (Fig. 2), the structure of the flattened area consisted of coarse acicular martensite with a small amount of pearlite (quench troostite) and ferrite (Fig. 3). At the surface and to a depth of 1.5 mm the case layer was coarse-acicular martensite and contained considerable amounts of retained austenite (Fig. 4). Therefore the sleeve was overheated and quenched probably directly from case.

Figure 5 shows the measured hardness values. The surface hardness is affected by retained austenite formation as had been determined already metallographically, and the core hardness at the flattened points is too high for a surface hardened piece that should have a core of some strain capacity.

It is understandable that a work piece with such a heterogeneous structure must be under high intrinsic stress caused by the differences in time sequence of cooling and transformation, and that these stresses lead or contribute to cracking during hardening or subsequent operation. The fracture cause therefore is too deep a carburization and

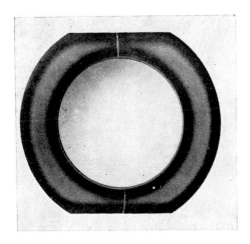

Fig. 1. Transverse section through cracked part of sleeve, etch: 5 % Nital. 1 x

Fig. 2. Cylindrical part

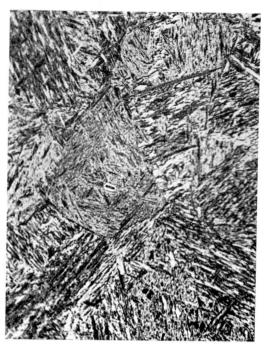

Fig. 4. Peripheral structure in 1.5 mm depth, transverse section, etch: 1 % Nital. 500 x

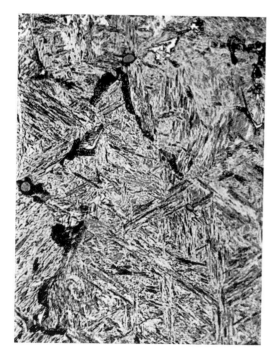

Fig. 3. Flattened part

Figs. 2 and 3. Grain structure, transverse section, etch: 1 % Nital. 500 x

hardening which prevented the reduction of stress by creep of the core material. Case depth is deeper than normal; but even if it would correspond to the specification which states that a hardness of at least 55 RC is necessary at a depth of 1.5 mm — calling for a carbon content of 0.4 to 0.5 % — only a narrow carbon-deficient core would remain. Therefore the mistake occurred already in the planning stage.

To prevent damage, it would have been necessary to

1) have a lower case depth,

2) carburize less deeply, and

3) prevent overheating that causes brittleness and leads also to increased case depth, or else use a fine grained steel of lower hardenability.

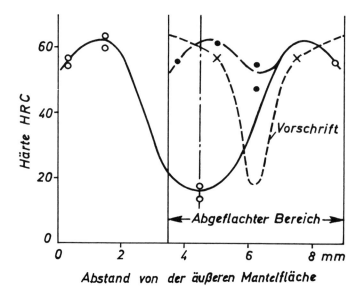

Härte = hardness
Vorschrift = specification
abgeflachter Bereich = flattened region
Abstand . . . = distance from outer
sleeve surface

Fig. 5. Hardness curve for wall in cylindrical and flattened region

154

Brittle Zinc Layer on a
Hot-Galvanised Hook

Egon Kauczor
Staatliches Materialprüfungsamt an der Fachhochschule
Hamburg

The surface of the hook did not possess the smooth and shiny zinc bloom surface normally observed on hot galvanised steel parts, but was matt and rough. Figure 1 shows part of the hook surface as received. Large cracks can be observed in the zinc layer.

A specimen was taken from a region at which the zinc layer was still adhering for metallographic micro-examination. Figure 2 shows the microstructure of the zinc in a section taken perpendicular to the surface. It can be seen that in the approx. 400 μm thick zinc layer, the pure zinc layer has been consumed by strong alloy formation. Such

zinc layers are brittle and can crack and peel off under mechanical stress if the layer is more than 300 μm thick [1]).

When the material to be galvanised is immersed in the zinc bath at approx. 450° C, the zinc alloys with the iron in the steel, assuming a perfectly clean steel surface. Alloy layers form one upon the other with average iron contents of 25 % (gamma layer) 9 % (delta layer) and 6 % (zeta layer). As the work piece is removed, a pure zinc layer forms on top of these (eta layer), which possesses a shiny surface and the well known zinc bloom. In the case of normally

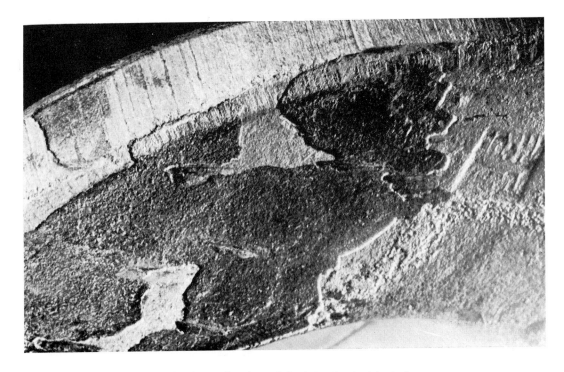

Fig. 1. Sectional photograph of the damaged surface of the hot galvanised hook. 2 x

Fig. 2. Microstructure of the zinc layer on a section taken perpendicular to the surface of the hook at a point at which the zinc layer was still adhering. The pure zinc layer has been consumed by excessive alloy formation. 200 x

galvanised objects the gamma layer is very thin and therefore frequently invisible in the microsection. In the case of parts which have had prolonged contact with the molten zinc, this phase, as in Fig. 2, is clearly visible in the extremely thick zinc coating. As comparison, Fig. 3 illustrates an example of a zinc layer of the usual thickness with normal structure.

The hook was made of silicon-killed alloy steel 41 Cr 4. A silicon content of 0.27 % was established analytically. Silicon accelerates the reaction between iron and zinc, which should have been taken into account in the present case by reducing the dip time or a small addition of aluminium (0.1 to 0.2 %) to the galvanising bath in order to retard the extremely rapid growth of the zinc layer and the strong alloy formation.

Even in the case of steel parts with lower silicon contents the reaction between iron and zinc can continue until the pure zinc layer has been consumed entirely if the work piece is not cooled sufficiently rapidly after withdrawal. The zinc coating then con-

sists of only the iron-zinc alloy layers. Since the grains of the iron-zinc phases grow in different directions and at different rates the surface becomes rough and matt [2]).

This phenomenon of "heat spot" appears when the galvanised material takes up too much heat from an overheated zinc bath. The same effect can also appear when the zinc bath is at the correct temperature if the parts have high mass : surface area ratio and, on account of their large heat capacity, remain long enough after withdrawal at sufficiently high temperatures to maintain the iron-zinc reaction until the pure zinc layer has been completely consumed. The rate of cooling of the still hot, galvanised components can be reduced critically by stacking. Accelerated cooling by blowing with compressed air or in the case of uncomplicated components by cooling in water retards the reaction in the pure zinc layer.

Heat spot can be brought about deliberately in hot galvanising by supplying heat directly to the galvanised material after withdrawal

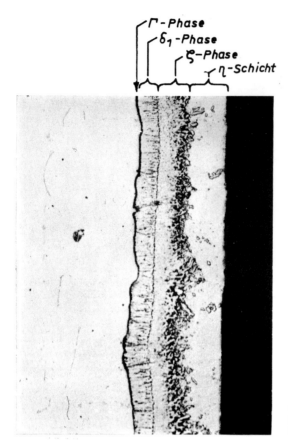

Γ-Phase
δ₁-Phase
ς-Phase
η-Schicht

Fig. 3. Comparative micrograph showing a zinc layer of the usual thickness with normal structure (controlled galvanisation). The alloy layers are covered by a pure zinc layer with only isolated embedded alloy grains. 200 x

(zinc annealing) until the pure zinc layer has been consumed. Paints and varnishes adhere better to the rough surface. In order to avoid cracking care must be taken that the zinc layers formed in this case are not too thick.

Literatur/References

1) J. F. H. VAN EIJNSBERGEN, Der Oberflächenschutz durch Feuerverzinkung an Stahlbauteilen in Industrieluft, Metall (1971) 985/992

2) H. BABLIK, Ungeeigneter Eisenuntergrund für das Feuerverzinken, Metalloberfläche (1950) A49/A53

Dimpling of Aluminium Sheet

Ruth Wittig and Jürgen Rickel
Vereinigte Flugtechnische Werke-Fokker G.m.b.H.
Bremen

Countersunk riveted joints in aluminium sheet are widely employed in the aircraft industry. The preparation of the sheet for the riveting process consists either of countersinking where the sheet is sufficiently thick or of dimpling. The dimples can either be pressed (pulled through) or stamped. Stamped dimples are preferred because they require a considerably smaller time expenditure than pressed.

In order to check the dimple stamping machines and the dimpling process, dimpled specimens are produced which can then be tested mechanically and metallographically.

The metallographic assessment of dimple defects will be described in the following. The specimens are made of clad aluminium sheet of alloy type Al Zn Mg Cu 1.5.

Figure 1 is a schematic representation of the mode of operation of a dimpling machine and Fig. 2 a perfect dimple in aluminium sheet for countersunk rivets and screws. Figures 3a and b illustrate a dimple with partially missing stamped surface. Missing stamped surface is known as „bell mouth" and arises if the stamping cylinder force is too low. Here a piece of the squeezed out

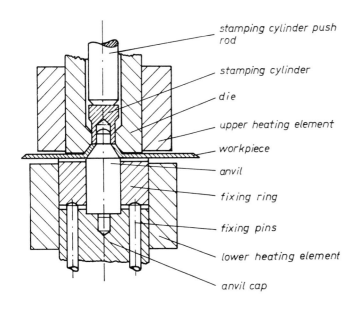

stamping cylinder push rod

stamping cylinder

die

upper heating element

workpiece

anvil

fixing ring

fixing pins

lower heating element

anvil cap

Fig. 1. Schematic representation of the mode of operation of a dimpling machine

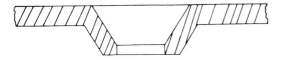

Fig. 2. Perfectly shaped dimple. The flattening of 50 % of the conical wall thickness is necessary only with aluminium sheet

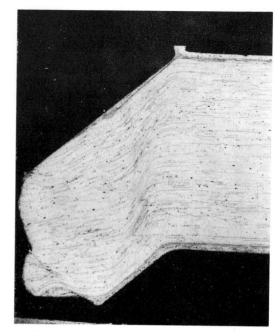

Fig. 3 a. 4 x

Fig. 3 b. 30 x

Figs 3 a and b. Dimple with partially missing stamped surface: "Bell mouth". The flashes are the residual stamped surface.

Fig. 4 a. 4 x

Fig. 4 b. 30 x

Figs. 4 a and b. Dimple with cylindrical protrusion

Fig. 5 a. 4 x

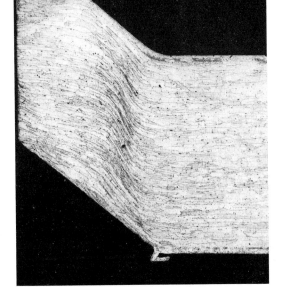

Fig. 5 b. 30 x

Figs. 5 a and b. Dimple with annular flash at the top surface of the sheet; "Ringed dimple" and flashes on stamped surface.

stamped surface has remained hanging on the bell mouth (Fig. 3b).

Figures 4a and b show a good stamping which however has been pressed out in the form of a cylindrical prominence. In this case the dimpling force was too great and the stamping cylinder force too low. The cladding was cut by the edge of the anvil and thrown up as the anvil was retracted.

Figures 5a and b show a dimple with flashes at the top surfaces of the sheet as a result of play between the stamping cylinder and the anvil head, a so-called „ringed dimple." There are also flashes on the otherwise satisfactory stamped surface.

Frequently, overlapping of several defects occurs, especially with steel or titanium sheet, with the result that it is difficult to identify the defects. The particularly well-pronounced defects illustrated here are intended to facilitate the assessment of dimples.

Cast Steel Housing with Grain Boundary Precipitates

Friedrich Karl Naumann and Ferdinand Spies

Max-Planck-Institut für Eisenforschung

Düsseldorf

In a housing made of cast steel GS 20 MoV 12 3, weighing 42 tons, precipitates were found on the austenitic grain boundaries during metallographic inspection. It was suspected that these consisted of aluminum nitride such as were previously observed repeatedly in the primary or austenitic grain boundaries of the structures of castings that had conchoidal tensile fractures, and displayed unsatisfactory properties [1]. The Institute was asked to test them. The steel for the casting in question had been deoxidized in the laddle with 0.29 kg aluminum per ton of raw steel. The casting had been cooled to 300° C in the mold, which took approximately two weeks. Of this time the cooling from 1200° to 900° C, i. e. through the particularly dangerous region of conchoidal fracture formation took 50 hours alone. After heat treatment in air the mechanical properties were passable. Nor did the tensile specimens show conchoidal fractures. Nevertheless, an investigation to clarify the microstructure phenomenon appeared rewarding.

The examined specimen had the following composition:

C %	Si %	Mn %	P %	S %
0,19	0,48	0,85	0,012	0,006

Cr %	Mo %	V %	Al %	N %
0,16	1,16	0,39	0,012	0,0095

Nitrogen and aluminum concentrations and the product $Al \cdot N$ were so low that on the basis of prior experience [1] a tendency to conchoidal fracture could not be expected.

Metallographic examination confirmed the presence of precipitates on the austenitic grain boundaries (Fig. 1). According to their shape and type they were recognized as carbides that were precipitated during tempering. In addition a much coarser network of rod-shaped and plate-shaped precipitates was found, that probably corresponded to the primary grain boundaries or to the grain boundaries or twin planes of the austenite formed during solidification of the melt. These particles could have been aluminum nitride judging by their shape and order of precipitation.

After heat treatment for greater toughness and quenching from 930° C in water followed by tempering to 600 ° C the specimen showed in part a coarse-conchoidal fracture structure (Fig. 2).

Plate-like, transparent particles could be isolated (Fig. 3) from the conchoidal fracture plane after carbon vapor deposition and electrolytic stripping of the films. They had the structure and the lattice constants of vanadium carbide as could be seen in the electron diffraction pattern. Particles of the same structure could also be extracted with lacquer films from a deeply etched section (Fig. 4). The section specimen had previously been quenched from 930° C in water in order to eliminate misleading effects of other carbides. During this operation the precipitates on the austenitic grain boundaries went

[1] F. K. NAUMANN, E. HENGLER, Stahl u. Eisen 82 (1962) 612/621

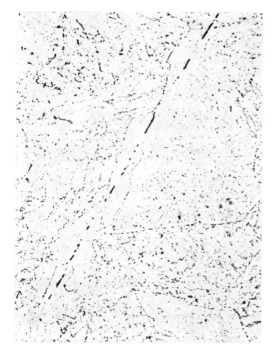

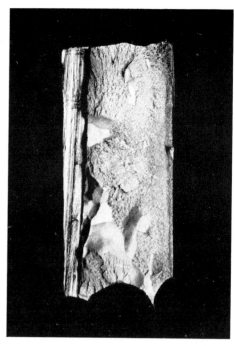

Fig. 1. Microstructure of casting, etch: Picral, 500 x

Fig. 2. Fracture after heat treatment 930°C/W., 600° C/ A. C. 1 x

Fig. 3. Precipitates stripped from fracture plane. 20 000 x

Fig. 4. Vanadium carbide extracted from section surface. 40 000 x

into solution. This confirmed the judgement that they were carbides precipitated during tempering.

Further annealing tests showed that the vanadium carbide dissolved on the primary grain boundaries only at substantially higher temperatures. Annealing for 2 hours at 1000°C decreased their quantity and the proportion of the conchoidal fracture of the structure very little; even after annealing for the same length of time at 1100° C traces were still present. Only after annealing at 1200° C was it possible to bring the carbides completely into solution and to eliminate the conchoidal fracture entirely.

It remained to be established at what temperature the dissolved carbide would be re-precipitated during slow cooling. For this purpose specimens were annealed for five hours at 1200° C and were cooled to 900° C by 4 to 1° C per minute. The cooling cycle was interrupted for 24 hour periods each at 100° C intervals. From 1000° C downward vanadium carbide again precipitated in in-creasing amounts as the temperature de-creased. Remarkably this did not occur at the boundaries of the newly formed austeni-tic grains but apparently at the undissolved nuclei of the original location.

According to these tests a subsequent re-moval of this defect by solutioning is im-practical because the annealing tempera-ture is too high. In order to avoid this defect in the future the sole recommendation to be made is to accelerate the cooling rate through the critical region between 1200 to 900° C to such an extent as is practicable with respect to machinability.

Destroyed Screen Bars of Stainless Steel

Friedrich Karl Naumann and Ferdinand Spies

Max-Planck-Institut für Eisenforschung

Düsseldorf

Screens made of stainless chromium-nickel-molybdenum steel X 5 CrNiMo 18 10 (Material No. 1. 4401), which were exposed to cooling water from the mouth of a river, became unserviceable after a few months because of the breaking out of parts of the bars. The bars of which the screen was assembled consisted of wires of wedge-shaped cross section with pressed-in ribs (Fig. 1). They were bent into loops at approx. 100 mm intervals and cold-pressed into final shape in two operations. With the aid of the loops they were then placed on rods. Short lengths of tubing were employed to maintain a distance of 8 mm between them. The rows of loops were welded after bracing. The cooling water contained 1.3 to 1.7 % Cl' and about 0.2 % SO_4".

A series of specimens which were ruptured on one or both sides was investigated. In all cases except one, in which a welding globule constituted the origin of the fracture (Figs. 2 and 3), the ruptures originated at a loop; specifically near the top at the contact surface of the turns (Fig. 4). This surface was strongly attacked by corrosion in all loops (Fig. 5). The contact surfaces between loops and spacers had also begun to rust. Frequently short surface cracks were observed in the fiber which had been under tensile stress during the bending of the loops (Fig. 6). One loop was even broken twice, namely once at the usual location, and again at an unusual location in the middle of the turn.

Here also a welding globule was later discovered at the origin of the fracture.

All specimens were investigated metallographically in longitudinal and transverse sections. Figure 7 is a transverse section, showing crevice or contact corrosion on a contact surface of the turns. Fig. 8 shows a longitudinal section of a fracture and next to it a starting crack. The longitudinal section shown in Figs. 9 a and b, reproduces a location close to the previously mentioned

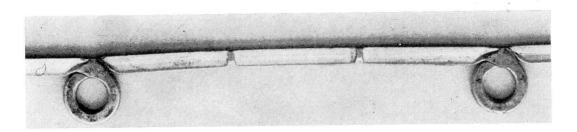

Fig. 1. Unused screen bar. 1 x

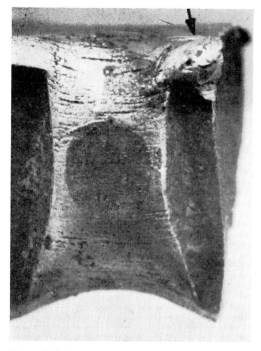

Fig. 2. Sideview

Fig. 3. Fracture

Figs. 2 and 3. Welding globule (arrow) at origin of fracture. 20 x

second fracture in the loop of a bar originating at a welding globule. The phenomena are to be interpreted as pitting and stress-corrosion cracking. An indication of the origin of the stresses is given by the slip lines and deformation twins shown in the etched section in Fig. 9 b. Figure 10 shows the structure of the welding globule.

The multiple fracturing of the screen bars in the brackish water of the mouth of the river must therefore be attributed to stress corrosion and pitting. The steel used, which contained molybdenum, would have withstood the severe corrosive conditions in the heat-treated condition, i. e. quenched after high temperature anneal. However, the stresses caused by deformation and welding, as well as the intensification of corrosive conditions brought about by design i. e. creation of corrosion currents in the poorly aereated gaps (Evans elements), have made this impossible.

Fig. 4. Fracture in loop. 10 x

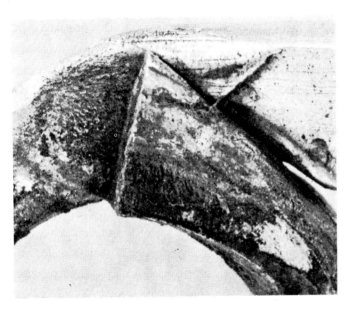

Fig. 5. Gap corrosion in loop. 10 x

Fig. 6. Surface cracks in loop. 10 x

Fig. 7. Corrosion at a contact surface of loop, transverse section, unetched. 100 x

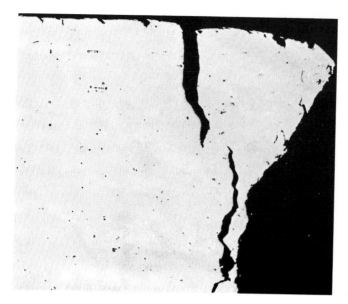

Fig. 8. Longitudinal section through fracture (right) and starting crack, unetched. 100 x

Fig. 9 a. Unetched. 200 x

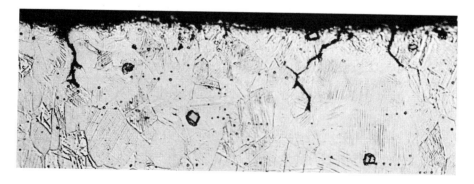

Fig. 9 b. Etched: V2A-etching solution. 500 x

Figs. 9 a and b: Pitting and stress-corrosion cracks, next to the fracture, longitudinal section

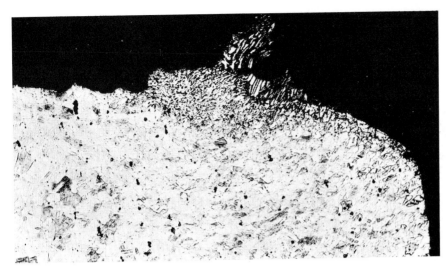

Fig. 10. Welding globule at origin of fracture, longitudinal section, etched with V2A-etching solution. 100 x

Wire Fractures in Suspension Cables of a Twisted Cable Suspension Bridge

Friedrich Karl Naumann and Ferdinand Spies

Max-Planck-Institut für Eisenforschung
Düsseldorf

During construction of a river bridge with 80 twisted cables, one or more fractures were found in each of 21 wires of 18 cables before assembly. All were located at the outside wrapping whose Z-profile wires were galvanically zinc-coated.

Four fractured and nine non-fractured wires of the outside wrappings and 17 non-fractured ungalvanized wires of the inside wrappings were examined macroscopically and microscopically. The fractures originated in all cases at the edge of the Z-profile (Fig. 1). These edges showed after dissolution of the zinc layer openings (Figs. 2 and 3) in the inside as well as in the outside wires. These fissures seemed to be deeper in the fractured than in the non-fractured wires.

But the decisive difference lay in the shape

of the cracks as could be seen in the longitudinal section through the edges of the wire profile. In the non-cracked and the ungalvanized wires of the inside wrappings they resembled the twisted flutes that coincided in part with the grain boundary ferrite of the decarburized peripheral layer.

Therefore they could have appeared originally during rolling (Fig. 4)[1] even though this could not be proved anymore after pickling and drawing. Experience shows that defects of this kind and depth have only a barely noticeable effect on strength and a very limited effect on pliability of the wire as could be confirmed by tests. Therefore, it is improbable that they were exclusively responsible for the wire fractures.

However, it could be seen repeatedly in the

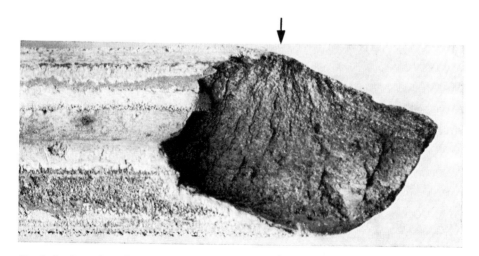

Fig. 1. Fracture of a galvanized outer wire, fracture origin designated by arrow. 5 x

Fig. 2. Fracture edge of wire according to Fig. 1, dezincified. 10 x

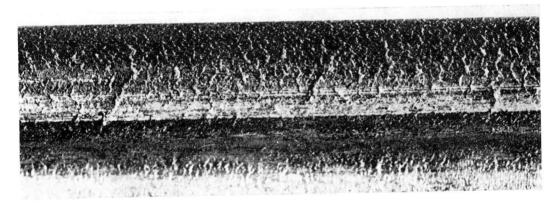

Fig. 3. Same edge of an ungalvanized inner wire. 10 x

Fig. 4. Longitudinal section through edge of Fig. 3, unetched. 100 x

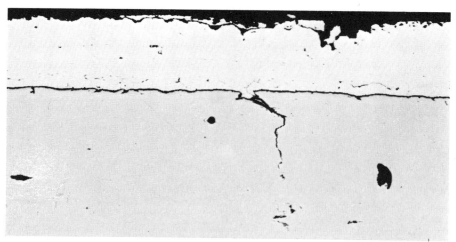

Fig. 5. Fissure with contiguous stress crack in fractured galvanized wire. Longitudinal section through edge from which fracture propagated, unetched. 100 x

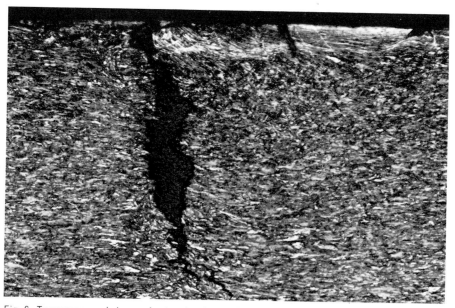

Fig. 6. Transverse crack in non-fractured galvanized wire, longitudinal section through fissured edge, etch: Picral. 100 x

cracked wires that a jagged crack was immediately contiguous to the fissure of the twist, and it ran almost vertically to the interior of the wire. This jagged crack ran in a transverse direction to the fiber and had the characteristics of a stress crack (Figs. 5 and 6). Therefore the suspicion arises, although this cannot be proved, that hydrogen played a role during crack formation, and that it penetrated during pickling or galvanizing. This supposition [2]) is confirmed also by the fact that the wire fractures were not observed during cable winding, but only subsequently to it, and therefore seemed to have appeared only after a certain delay.

Literatur/References

[1]) vgl. P. FUNKE, H. KRAUTMACHER, R. OHLER, Stahl u. Eisen 87 (1967) 318/331
[2]) vgl. F. K. NAUMANN, F. SPIES, Prakt. Metallographie 8 (1971) 724/728

Broken Helical Compression Spring

Friedrich Karl Naumann and Ferdinand Spies

Max-Planck-Institut für Eisenforschung

Düsseldorf

A helical compression spring with 10 turns made of 1.8 mm thick wire which was under high pressure during tension applied to a rocker arm broke on the test stand in the third turn.

The fracture was a torsion fracture that initiated in the highly loaded inner fiber and showed in its origin the characteristics of a fatigue fracture (Fig. 1). A longitudinal fold was located at the fracture crack break-through (Fig. 2) which could still be observed at the fourth and fifth turns, where a further incipient crack originated (Fig. 3).

A metallographic section was made directly next to the fracture path and the fold was cut (Fig. 4). It showed decarburized edges in

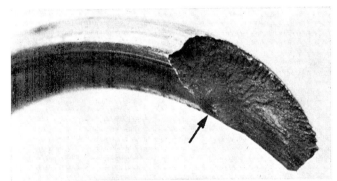

Fig. 1. Fracture of spring, arrow = fracture path. 10 x

Fig. 2. Inner view of third turn with fracture path (arrow). 10 x

Fig. 3. Incipient fracture of fourth turn. 10 x

the outer slanted part and thus most likely occurred during rolling (Fig. 5). The inner radially proceeding part, however, was probably a fatigue fracture originating in the fold.

The fracture of this highly stressed spring was therefore accelerated by a rolling defect. In order to decrease the stress, the construction has meantime been modified.

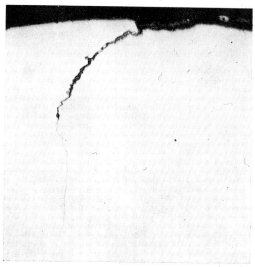

Fig. 4. Unetched

Fig. 5. Etched with Picral

Figs. 4 and 5. Cross section immediately next to fracture. 200 x

Ruptured Prestressing Cables from a Viaduct

Friedrich Karl Naumann and Ferdinand Spies

Max-Planck-Institut für Eisenforschung
Düsseldorf

During the construction of a prestressed concrete viaduct, several wires 12.2 mm in diameter ruptured after tensioning but before the channels were grouted. They were made of heat treated prestressed concrete steel St 145/160 with a minimum yield stress of 145 kgf/mm² and tensile strength of 160 kgf/mm². While the wire bundles, each containing over 100 wires, were being drawn into the channels they were repeatedly pulled over the sharp edges of square section guide blocks.

Two short pieces of cable with fractures and five ½ m long pieces without were sent for examination. They exhibited localised shiny or bluish tinged chafe zones on the surface (Fig. 1) which in some cases were cracked or could be made to crack by bending (Fig. 2).

The fractures were initiated at these chafe zones. The fracture surfaces showed a kidney shaped crack darkened by corrosion and therefore old and a large catastrophic fracture with deformed border (Fig. 3).

A longitudinal section through the origin of fracture was taken in the case of both the wires with fractures. Figure 4 shows the microstructure of the outer zone of one of the wires adjacent to the incipient fracture. It shows heavy deformation and tonguelike overlaps. The material has become so hot as a result of deformation that martensite has formed as it cooled rapidly. The same phenomena were observed on the second wire. In this case there was a clear association between the fracture and a transverse crack in the martensitic outer layer (Fig. 5).

The long wires without fractures also showed the same structural changes at the surface though not as well pronounced as in the short fracture specimens. Figure 6 shows the microstructure of such a specimen in which the martensite layer has been cracked by stretching.

In tensile tests, the long specimens reached

Fig. 1. Chafe mark on the cable surface. 6 x

Fig. 2. Cracks in a chafing zone occurring on bending. 20 x

Fig. 3 a

Fig. 3 b

Figs. 3a and b. Fractured wires. 4 x

yield stresses of 150 to 154 kgf/mm² and strengths of 165 to 167 kgf/mm² thus satisfying the specifications. They all necked strongly before fracturing. In two specimens the cup and cone fracture was initiated from within (Fig. 7) but in three from chafe marks on the outer surface (Fig. 8). Compared with the former, the latter showed a reduction in strain from 7 to 6 % ($L_0 = 10$ d) and a reduction in necking from 43 to 25 %.

It can be concluded from these findings that the chafing of the wires on the edges of the guide blocks, particularly the resulting martensite formation, caused the wires to rupture. The martensite formation may well have led to cracks in the structure as a result of transformation stresses or the brittle martensite may have been cracked by localised plastic deformation of the wire. In any case the cracks act as notches which, particularly when corrosion takes place, can lead to fracture even when the stress lies well below the yield point. The error here lay more in the planning than in the design.

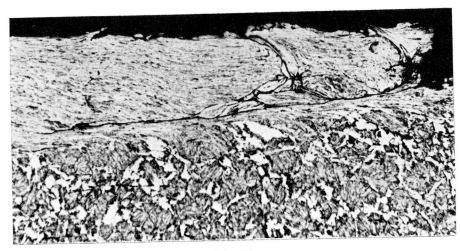

Fig. 4. Chafing zone adjacent to fracture origin. 500 x

Fig. 5. Chafing zone with fracture origin (arrow). 200 x

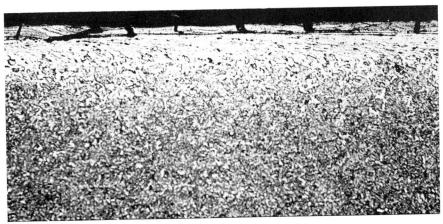

Fig. 6. Chafing zone on a cable with no fractures (cf. Fig. 2). 100 x

Figs. 4 to 6. Longitudinal sections etched in picral

(Continued on the next page)

Fig. 7. Cup and cone fracture originating from within the wire (44 % necking)

Fig. 8. Fracture originating at a chafe mark (19 % necking)

Figs. 7 and 8. Fractured tensile specimens. 4 x

Broken Full Lift Disk of a Safety Valve

Egon Kauczor

Staatliches Materialprüfungsamt an der Fachhochschule
Hamburg

The full lift disk, made of die cast brass, which served as a lifting aid in a safety valve, had cracked in service at a number of locations in the vicinity of the threaded hole. A photograph of the overall view, shown in Fig. 1, reproduces part of the surface in the as-received condition.

A specimen removed perpendicular to the surface for metallographic preparation shows numerous pores recognizable with the unaided eye (Fig. 2).

During the microscopic examination, agglomeration of oxide inclusions were noted in the region of the cracks. Since the die cast brass under consideration was alloyed with aluminium, these inclusions consisted predominantly of aluminium oxide (Fig. 3).

Pores and oxide inclusions are hardly unavoidable in die cast components. This is because during the discharge of the shot, air is often entrapped in the pressure chamber and forced into the die. The more com-

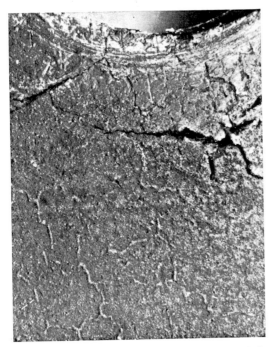

Fig. 1. Photograph showing overall view of part of the surface of the fractured lift disk in the as-received condition. 6 x

Fig. 2. Macroscopically visible pores in a section perpendicular to the surface. 8 x

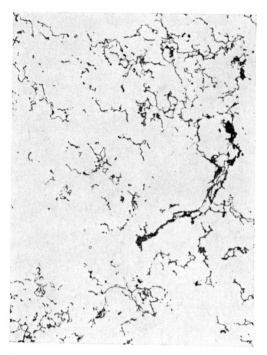

Fig. 3. Agglomeration of oxide inclusions in the region of the cracks. Unetched microsection. 500 x

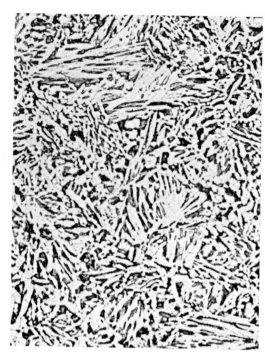

Fig. 4. Microstructure of a sound region. Etched with ferric chloride. 500 x

plex the shape of the casting, the more difficult it becomes to avoid defects during die casting. This is because more frequent changes in the flow direction cause the liquid metal stream to become more turbulent and to include more air bubbles and oxides. For this reason usually only components which are not subjected to very great demands with regard to soundness under pressure are die cast [1]).

The tolerable limit in pores and oxide inclusions is greatly exceeded in the lift disk under examination. Above all, the numerous oxide skins disrupt the cohesion of the microstructure and are therefore to be considered primarily responsible for the failure of the lift disk.

The photomicrograph of a sound region (Fig. 4) shows a fine grained acicular $\alpha + \beta$

structure in the etched condition as is usual for brass die castings.

In addition to the cracks, Fig. 1 shows irregular lines on the surface which are impressions of hot cracks in the steel die. The elevated casting temperatures place such severe demands on the die material, that hot cracks are frequently formed before the minimum number of pieces (about 2000) is reached. This minimum is of necessity high in view of the costs of the dies [2]). These impressions are only to be viewed as surface blemishes which had no effect on the occurrence of fracture.

Literatur/References

[1]) G. OREHOUNIG, Luft- und Oxydationseinschlüsse in Messing-Druckgußerzeugnissen, Gießerei-Praxis, Heft 7 (1966) 109/113

[2]) K. DIES, Kupfer und Kupferlegierungen in der Technik, Springer-Verlag, Berlin, Heidelberg, New York, (1967)

Grain Disintegration in a Welded Sheet Construction

Egon Kauczor

Staatliches Materialprüfungsamt an der Fachhochschule
Hamburg

The object under investigation was the corner of a welded sheet construction made from austenitic corrosion-resistant chromium-nickel steel. Examination of the object as received showed corrosive attack of the outer sheet (Fig. 1), this attack being most severe at the points subject to the greatest heat during welding indicated by ↘. A photograph of the inner surface of the specimen, Fig. 2, shows that particularly large amounts of weld metal were applied here.

A sample was taken for metallographic examination from the point marked ·—→ ←·· in Fig. 1 for observation under the microscope. Figure 3 shows the unetched metallographically prepared surface. It can be seen that in the lower sheet the steel in the limited region of the heat affected zone of the welded seam has been attacked from both sides. The sectional micrograph in Fig. 4 clearly shows the grain boundary break-up typical of grain disintegration. The intergranular cohesion of the structure was so strongly affected that numerous grains broke out during polishing.

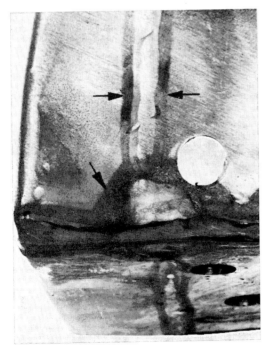

Fig. 1. Exterior

Fig. 2. Interior

Figs. 1 and 2. General views of the specimen as submitted. 1 x

Fig. 3. Unetched polished section taken from ·⟶ ⟵· in Fig. 1. 10 x

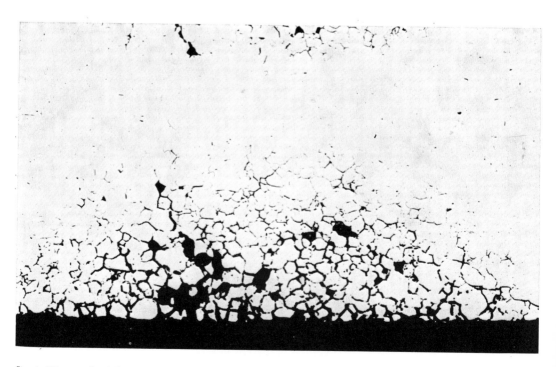

Fig. 4. Micrograph of the corrosion zone on the right hand side of Fig. 3 at higher magnification. 50 x

Spectral analysis revealed that the sheets were made of unstabilised chromium-nickel steels. The chemical analysis for carbon yielded 0.08 % C in the outer sheet and 0.05 % C in the inner sheet. Steels with more than 0.07 % C are not suitable for welding if the carbon is not bound to a stabilising element such as titanium or niobium.

If such steels are heated to temperatures between 425 and 870° C [1] the carbon comes out of supersaturated solution, particularly between 600 and 700° C, and forms carbides with a high chromium content which precipitate preferentially at grain boundaries. The matrix in the vicinity of the grain boundaries can become so depleted in chromium that it loses its passivity [2]. Corrosive agents coming into contact with a corrosion resistant steel rendered susceptible (sensitive) by chromium depletion eat into the steel along the grain boundaries (intergranularly). During welding, the critical temperatures are reached in the heat affected zone parallel to and at certain distances from the welded seam.

The best known measures for combating grain disintegration are the use of particularly low carbon steels or those in which the carbon is stably bound by additions of titanium or niobium.

In the present case grain disintegration has been promoted by the thickness of the bead and the consequently large quantity of heat required.

If non-stabilised austenitic sheet is to be used in future for this construction, it is recommended that one of the particularly low carbon steels X2 CrNi 18 9 or X2 CrNi Mo 18 10 be used which with a maximum carbon content of only 0.03 % are more resistant to grain disintegration.

Literatur/References

[1] A. L. PHILLIPS, Welding of Austenitic Chromium-Nickel Stainless Steels. Welding Handbook, 4. Ausgabe, Kapitel 65, American Welding Society, New York.

[2] B. STRAUSS, H. SCHOTTKY, J. HINNÜBER, Z. anorg. allg. Chem. 188 (1930) 309/324

Fractured Recuperator Made of Heat Resistant Cast Steel

Friedrich Karl Naumann and Ferdinand Spies

Max-Planck-Institut für Eisenforschung

Düsseldorf

A recuperator for blast heating of a cupola furnace became unserviceable because of the brittle fracture of several finned tubes made of heat resistant cast steel containing 1.4 % C, 2.3 % Si and 28 % Cr. The service temperature was reported as 850° C. This already led to the suspicion that the fracturing had something to do with the precipitation of sigma phase. The sigma phase consists of an intermetallic compound of approximately 50 at. % Cr which forms from the solid solution during slow cooling or during extended annealing below 800° to 900° C (depending upon the other components of the alloy). The formation of sigma phase is accompanied by a decrease in volume which puts the steel under stress [1]).

The fractures of the finned tubes were both coarse- and finegrained and undistorted which is not uncommon for this casting alloy (Fig. 1). For the metallographic investigation, in which was also included an unused recuperator for purposes of comparison the sections after the usual etching with V2A-Beize which only attacks the austenitic matrix, were first etched electrolytically in a 1:10 dilution of saturated ammonia water at a terminal voltage of 1.5 V. This served to dissolve the chromium carbide and color it by means of a brown precipitate. Immediately after and without intermediate polish, the section was etched at 2 V terminal potential in a 10n solution of caustic soda which in addition etched and colored the sigma phase. The result for the broken recuperator is reproduced in Figs. 2 b and c, and for the unused one in Figs. 3 b and c. Accordingly, the finned tubes of the unused

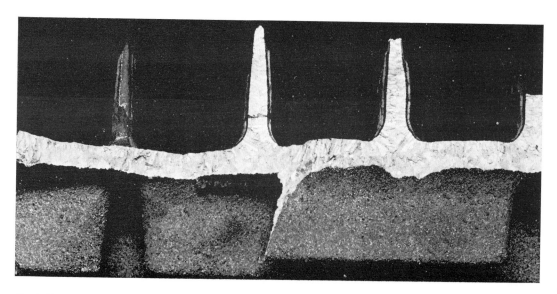

Fig. 1. Fracture of the finned tube. 1 x

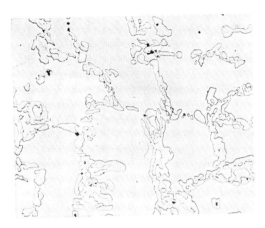 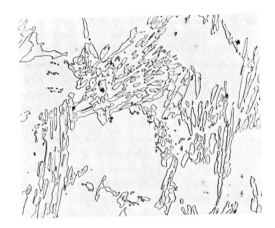

Figs. 2 a and 3 a. Etching treatment: V2A-etching solution

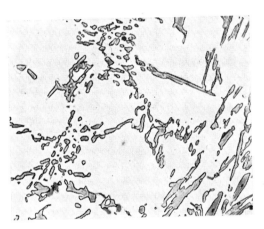 

Figs. 2 b and 3 b. Etching treatment: ammonia water, 1.5 V (etching of carbides)

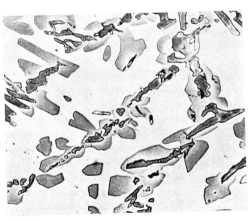 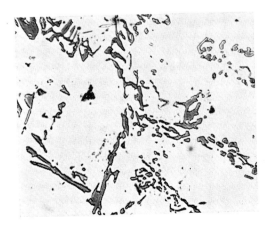

Figs. 2 c and 3 c. Etching treatment: ammonia water, 1.5 V + 10 n caustic soda, 1.5 V (etching of carbides and sigma phase)

Figs. 2 a to c. Microstructure of the fractured recuperator. 500 x

Figs. 3 a to c. Microstructure of the unused recuperator. 500 x

heat exchanger contained in addition to the chromium-silicon alloyed solid solution only chromium carbide as a precipitate; the caustic soda etch had no effect. On the other hand, in the fractured finned tubes, a considerable amount of sigma phase had formed as a result of the prolonged anneal at 850° C. In the photomicrograph this phase appears attacked and discolored by the second etch. The previously formed carbides served as nuclei for the precipitation of the sigma phase.

The multiaxial stresses caused by sigma phase formation and the related embrittlement must be viewed as the cause for the fracture of the recuperator. In this case the designer has evidently chosen an unsuitable construction material. A steel of lower chromium content with no or little tendency for sigma phase formation would have had adequate corrosion resistance at the relatively low service temperature.

Literatur/Reference

[1] F. K. NAUMANN, Arch. Eisenhüttenwes. 34 (1963) 187/194

Cracked Disks of Fan Made of Heat Resistant Steel

Friedrich Karl Naumann and Ferdinand Spies
Max-Planck-Institut für Eisenforschung
Düsseldorf

Three radially cracked disks that had served to circulate the protective gases in a bell-type annealing furnace were sent in for examination. During service they had been heated in cycles of 48 hours to 720°C for 3 hours each time, then were kept at temperature for 15 hours followed by cooling to 40°C in 30 hours, while rotating at 1750 rpm.

Two disks were cracked at the inner face of the sheet metal rim while the rim of the third was completely cracked through (Fig. 1). For metallographic investigation one of the cracks was ground flat in the sheet plane. It could be seen that the crack branched out intergranularly (Fig. 2). Beyond the end of the crack, grain boundary separations could be discerned (Fig. 3). The microstructure of the sheet consisted of austenite with strong precipitates in the grain and at the grain boundaries (Fig. 4). The precipitated phase was not attacked when etched electrolytically with ammonia water (Fig. 5), but was discolored brown by caustic soda solution (Fig. 6). Therefore it did not consist of chromium carbide but of σ-phase[1]. The fracture followed along the σ-precipitates at the grain boundaries (Fig. 7). The cracking of the disks therefore was caused or favored by σ-phase formation and stress concentrations and embrittlement connected with it.

Fig. 1. View of cracked disk. 1/5 ×

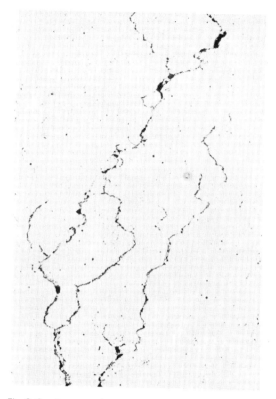

Fig. 2. Crack propagations in unetched flat cut. 50 ×

Fig. 3. Grain boundary separations outside of crack area. Unetched flat cut. 50 ×

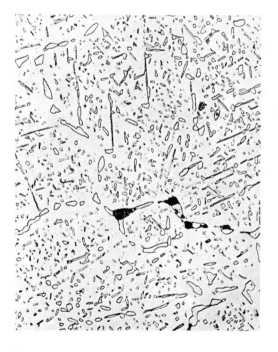

Fig. 4. Etch: V2A-pickle

Fig. 5. Electrolytic etch with ammoniawater, 1.5 V

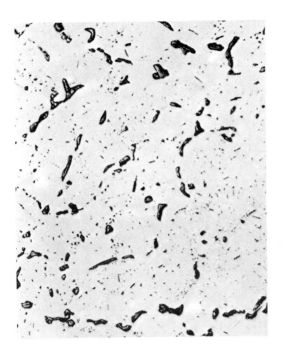

Fig. 6. Electrolytic etch with caustic soda 10 n, 1.5 V

Figs. 4 to 6. Microstructure of sheet. 500 ×

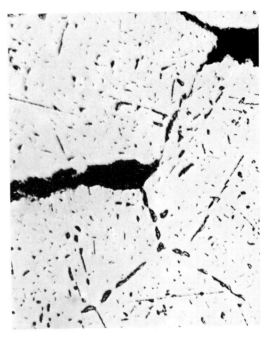

Fig. 7. Intergranular crack in sheet. Electrolytic etch with 10 n caustic soda, 1.5 v. 500 ×

An analysis of the sheet metal rim of one of the disks showed the following composition:

C %	Si %	Cr %	Ni %
0,06	1,98	25,8	35,8

A steel of such high chromium content is susceptible to σ-phase formation when annealed under 800° C[1]). The material selected was therefore unsuitable for the stress to be anticipated. In view of the required oxidation resistance, a chromium-silicon or chromium-aluminum steel with 6 or 13 % Cr would have been adequate (Materials Nos. 1.4712 or 1.4713, or 1.4722 or 1.4724, respectively). If the high temperature strength of these steels proved inadequate, an alloy lower in chromium, such as X 12 NiCrSi 36 16 with 16 % Cr (material No. 1.4864) would have been preferable.

[1] F. K. NAUMANN, Beitrag zum Nachweis der σ-Phase und zur Kinetik ihrer Bildung und Auflösung in Eisen-Chrom- und Eisen-Chrom-Nickel-Legierungen. Arch. Eisenhüttenwes. 34 (1963) 187/194

Worn Cast Iron Pump Parts

Friedrich Karl Naumann and Ferdinand Spies
Max-Planck-Institut für Eisenforschung
Düsseldorf

A slide and the two guideways of a pump had to be disassembled already during run-in time after approximately 20 h because they had galled completely, before the rated speed of 800 rpm was reached.

Figure 1 shows a section of the worn surfaces. For metallographic examination, transverse sections through the galled regions A--A, D--D and F--F and horizontal sections parallel to the galled surfaces of all three parts were made.

Figure 2 shows the structure under the gliding surface of the slide (Section A--A). It contains considerable amounts of ferrite. At the edges which form the gliding surface with the unworked windows (points C) the structure consists predominantly of granular graphite in a ferritic matrix (Fig. 3). The ferrite content is still higher in the steps between the windows (sections D--D and E--E), as well as in the center (Fig. 4), and especially in the vicinity of the edge (Fig. 5). In contrast to this, the structure of the gui-

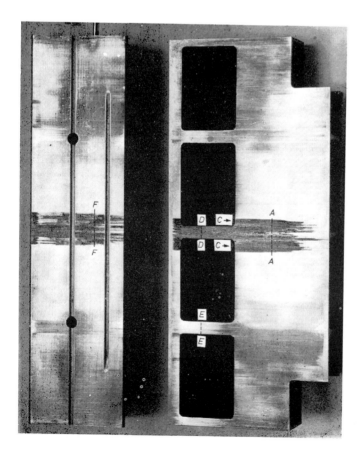

Fig. 1. Galled surfaces, left: guideway, right: slide. ¼ x

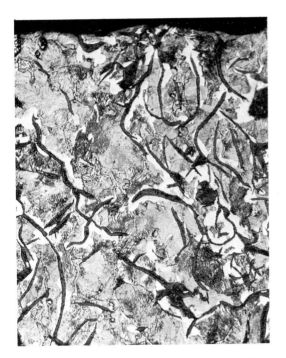

Fig. 2. Under gliding surface, transverse section A—A

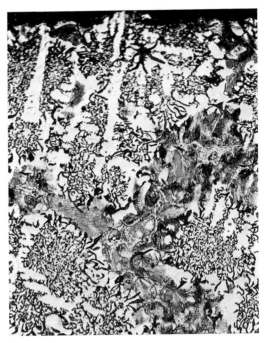

Fig. 3. At edge with casting skin, horizontal section, Point C

Fig. 4. In center of gliding surface, transverse sections D—D and E—E

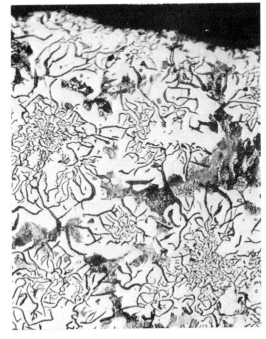

Fig. 5. At edges of gliding surface, transverse sections D—D and E—E

Figs. 2 to 5. Structure of slide, etch: Picral. 200 x

Fig. 6. Structure of guideway with thicker cross section in hemispherical part, section F––F, etch: Picral. 200 x

deways that had a considerably thicker cross section, is almost entirely pearlitic (Section F--F, Fig. 6).

The ferritic structure endows the parts with poor wear resistance and a tendency to gall. This, of course, also encompasses the opposite surface. The granular graphite in a ferritic matrix is formed by supercooling of the melt during solidification. It appears, therefore, in thin-walled places during fast cooling or in the vicinity of the edges. Cast iron with high carbon content and saturation value has a strong tendency to this kind of supercooling, especially if the iron contains few crystallization nuclei.

Chemical analysis of the slide showed the following composition:

This corresponds to a saturation value of 1.08 according to $\dfrac{C}{4.23 - 0.312 \cdot Si - 0.275 \cdot P}$. The iron was thus distinctly hypereutectic.

The galling of the pump parts therefore was favored by an unsuitable structure caused by improper composition and fast cooling. Distortion by casting stresses may have been contributory or may have played the principal part. In order to prevent a repitition, the use of hypoeutectic or eutectic iron, slower cooling of the casting, inocculation of the melt with finely powdered ferrosilicon and possibly rounding-off the edges or machining of the surfaces are recommended.

C %	Graphite %	Si %	Mn %	P %	S %
3,60	3,22	2,49	0,51	0,485	0,112

Section II:
Service-Related Failures

Cracked Pipe Elbow of a Hydraulic Installation

Friedrich Karl Naumann and Ferdinand Spies

Max-Planck-Institut für Eisenforschung

Düsseldorf

An elbow of 70 mm O. D. and 10 mm wall thickness made from St 35.29, and exposed to 315 atmospheres internal pressure in an oil hydraulic shear installation, cracked lengthwise after a short operating period. Since the stress would not have been sufficient to explain the fracture of this elbow under this pressure, an investigation was conducted to establish whether material or processing errors had occurred.

Figures 1 and 2 give the exterior and interior view of the elbow with the crack. It was already cross-sectioned upon receipt. The crack was located laterally in the inside of the bend.

A section at the existing parting line was taken for metallographic examination. Macro-etching and Baumann print showed that the material for the pipe was of uniform purity (Fig. 3). But after etching with Nital, zones of 10 to 15 mm width and 1 to 1.5 mm depth with modified structure could be seen towards the bend at both sides of the bend plane. They are designated with a and b in Fig. 4 and are marked by broken lines. The crack was located at the edge of one. From the microscopic examination it could be seen that the ferritic-pearlitic structure in these locations (Fig. 5) was very fine-grained and also showed other signs of fast cooling as compared to normally

Fig. 1. Pipe elbow with longitudinal crack (arrow). 1 x

formed structure of the core zone (Fig. 6). It is also possible that the pipe was resting on a cold plate during bending or that it came in touch with a cold tool. This apparently caused the strains at the transition to the cross-sectional part that had been cooled more slowly. The location of the crack at just this point gives rise to the conclusion that it was formed either by the sole or contributive effect of these stresses.

Fig. 2. Longitudinal crack. 4 x

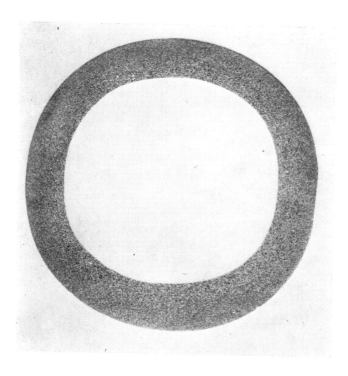

Fig. 3. Sulphur print according to Baumann

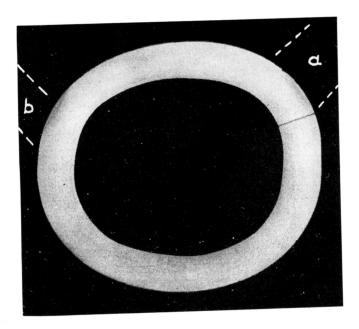

Fig. 4. Etch: Nital. approx. 1 x

Figs. 3 and 4. Cross section through center of elbow

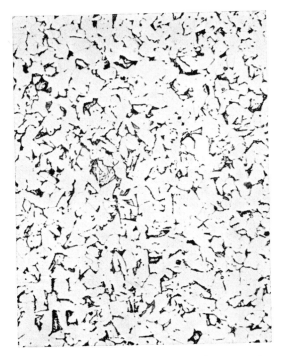

Fig. 5. Microstructure of locations a and b in Fig. 4

Fig. 6. Normal structure of pipe

Figs. 5 and 6. Transverse section according to Fig. 4. Etch: Picral. 500 x

Fractured Turbine Blades

Friedrich Karl Naumann and Ferdinand Spies
Max-Planck-Institut für Eisenforschung
Düsseldorf

In an electric power station, 7 turbine blades out of 112 blades, broke or cracked in a time period of 8 to 14 months after commencement of operation. The blades, in question were all located on the last running wheel in the low pressure section of a 35,000 kW high pressure condensing turbine. They were milled blades without binding wires and cover band. They did not fracture at the fastening, i. e. the location of highest bending stress, but in a central region which was 165 to 225 mm away from the gripped end. The blades were fabricated from a stainless heat-treatable chromium steel containing 0.2 % C and 13.9 % Cr.

Figure 1 reproduces the external part of a blade showing a crack (left side in picture). The surface was covered with a dark oily layer which was soluble in toluol and could be wiped off easily. The inlet edge showed strong erosion characteristics which increased towards the end of the blade (Fig. 2). The crack originated at this edge. Forcing the crack open revealed a fatigue fracture (Fig. 3). The fracture-surface was tarnished blue. Figure 4 reproduces the fracture of

another blade which had ruptured completely. This failure as well was initiated by a fatigue fracture starting from the eroded inlet edge.

Analysis confirmed that the blades consisted of a steel containing 0.2 % C and 13 % Cr as is customarily employed for such blades.

For the microstructure examination, longitudinal and transverse sections were taken from three blades in the proximity of the cracks and close to the completely worn-out blade end. They showed a normal heat treated structure (Fig. 5). The inlet edge was full of fissures and initial cracks (Figs. 6 and 7). The structure in this region showed signs of stress- or vibration-induced corrosion cracking. Near the blade ends the remains of a hardened zone were recognizable at the inlet edge (Fig. 8). The erosion was however not prevented by the hardening. The hardness (HV 10) amounted to between 245 and 255 kg/mm² corresponding to a tensile strength of about 84 to 87 kg/mm², and increased to approx. 500 kg/mm² in the hardened regions.

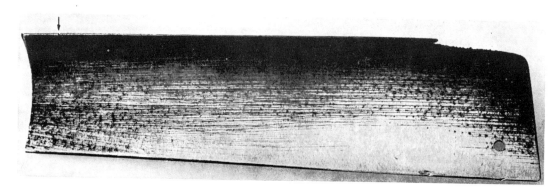

Fig. 1. Inner surface of the blade with crack. 0.5 x

The blades were therefore destroyed by flexural vibrations which evidently reached their maximum amplitude at the location of fracture. The erosion of the inlet edge, possibly in connection with vibration-induced corrosion cracking, contributed to fracture.

Fig. 2. Outer surface of the blade with erosion at the inlet edge and crack. 3 x

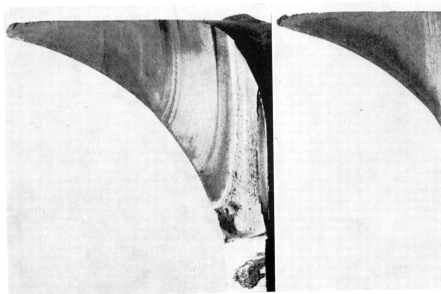

Fig. 3. Fracture surface of the crack which has been forced open. 2 x

Fig. 4. Fracture-surface at fractured end of blade stump. 2 x

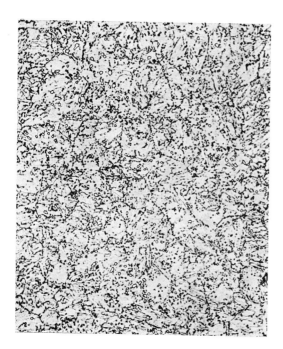

Fig. 5. Structure of the blade, transverse section, etch: V2A-solution. 500 x

Fig. 6. 200 x

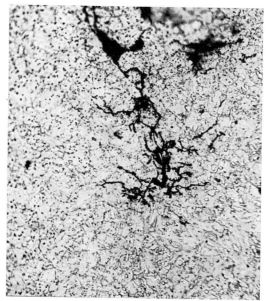

Fig. 7. 500 x

Figs. 6 and 7. Eroded inlet edge, transverse section, etch: V2A-solution

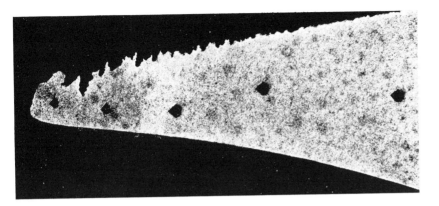

Fig. 8 a. Proximity of blade end

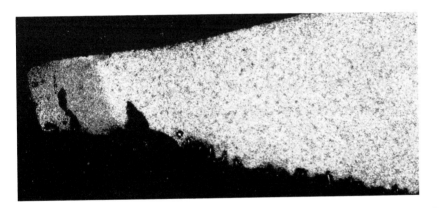

Fig. 8 b. Center of blade

Figs. 8 a and b. Eroded inlet edge with hardened region and Vickers hardness indentations, transverse section, etch: V2A-solution. 10 x

Cracked Slitting Saw Blades

Friedrich Karl Naumann and Ferdinand Spies

Max-Planck-Institut für Eisenforschung

Düsseldorf

Two slitting saw blades were delivered for the purpose of determining the cause of damage. One had cracked while the other one came from a prior sheet delivery, that had less tendency to crack formation according to the manufacturer. The blades were supposed to have been stamped out of a sheet made from a 55 kp/mm² strength steel. After heat treatment the cutting edge was hard chrome plated to increase wear resistance. The saw blades were used for separating steel profiles at high rotational speeds. The saw manufacturer assumed that the cracks had occurred already during stamping out of the blades.

As can be seen from Figs. 1 and 2 the cracks in question were located at the base of the teeth, i. e. at the point of highest operating stress. That contradicts the assumption that they may have been formed already during stamping. After breaking open a crack, an abraded fracture grain emerged as well as a dark stain (Fig. 3). The existence of a narrow grainy fracture zone around the center plane of the blade led to the assumption that several fractures had propagated also from the two lateral planes of the blade and met in the center. These were probably fatigue cracks that had occurred during the vibration of the blade.

Metallographic examination was made of sections of the blades that were cut perpendicular and parallel to the surface. Figs. 4 and 5 present peripheral microstructures of cross sections through the chrome plated zone below the tooth crown. In this zone, where incipient fractures that propagate from the circular planes also occur, quite a number of fine incipient cracks were intersected. It is well known [1]) that the electrolytic deposition of chromium layers promotes the formation of cracks and fatigue fractures due to the

Fig. 1. Defective saw blade with cracks at base of teeth (arrows)

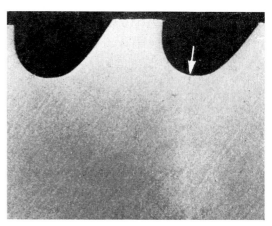

Fig. 2. Satisfactory comparison blade with short incipient crack at right tooth base (arrow)

Figs. 1 and 2. Lateral views. Surface etched with Nital. 3 x

Fig. 3. Opened crack (cutting edge on top). 3 x

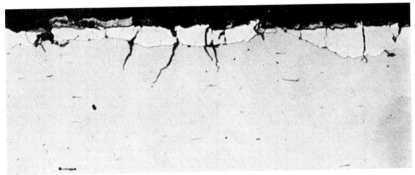

Fig. 4. Defective saw blade

Fig. 5. Satisfactory comparison blade

Figs. 4 and 5. Peripheral microstructure in chrome plated cutting edge. Transverse section, unetched. 500 x

Fig. 6. Transition structure under tooth backings of defective saw blade. Transverse section, etch: Nital. 500 x

Fig. 7. As Fig. 6. Pearlitic zone under martensitic peripheral microstructure. 500 x

Fig. 8. Heat treated microstructure of saw blades, transverse section, etch: Nital. 500 x

penetration of hydrogen into the steel, and the consequent build-up of internal stresses. The cracks shown in Figs. 4 and 5 therefore may be either hydrogen cracks or fatigue cracks. The satisfactory blade from the older delivery also showed these incipient cracks (Fig. 5) as well as some in the tooth base (Fig. 2). All cracks were non-decarburized and were free of chromium deposits. Therefore they could not have existed before heat treatment and chrome plating.

However, there was indeed a difference between the damaged saw blade and the one designated as satisfactory: the first mentioned showed a sickle-shaped zone with an altered microstructure under the back of the teeth. It was made visible in Fig. 1 by etching the surface with Nital, while it was absent in the satisfactory blade (Fig. 2). The zone consisted of a thin martensitic periphe-

ral layer (Fig. 6) that extended into the original heat treated structure (Fig. 8) by way of a zone of fine grained pearlite Fig. 7). Accordingly the tooth backings of the defective blade were heated to temperatures in the austenitic range during separation of the steel profiles and were transformed in part in the martensitic stage by fast cooling. This may have contributed to fracture of this blade through additional residual stresses.

Accordingly, the damage was due neither to poor quality of the sheet nor to defective stamping or heat treatment, but had occurred later either during surface treatment or during operation.

---

[1] H. WIEGAND, Metallwirtschaft 18 (1939) 83/85; vgl. a. F. K. NAUMANN, F. SPIES, Prakt. Metallographie 7 (1970) 707/712

Crankshaft with Torsion Fatigue Fractures in Inductively Surface-Hardened Crank Pin

Friedrich Karl Naumann and Ferdinand Spies

Max-Planck-Institut für Eisenforschung
Düsseldorf

A crankshaft which was overloaded on a test stand suffered an incipient crack in the crank pin (Fig. 1). The crack run generally parallel to the longitudinal axis and branched off at the entrance into the two fillets at the transition to the crank arm (Fig. 2 a). As shown in Fig. 2 b it consisted of many small cracks, all of which propagated at an angle of approximately 45° to the longitudinal axis, and therefore were caused by torsion stresses.

According to Fig. 2 a, a transverse section in the plane A – – A showed that the crank pin was surface hardened up to a depth of 2 to 2.5 mm and that the crack had penetrated the pin radially up to a depth of 14 mm (Fig. 3). The otherwise smooth crack had branched outward in the fine martensitic structure of the rim as well as in the bainitic structure of the core (Fig. 4).

Neither macroscopic nor microscopic examination could determine any material or processing faults in this connection.

The crack located at the left in section A – – A of Fig. 2 a of the pin section was broken open for further examination. The fracture followed the steps of the small incipient cracks at the surface (Figs. 5 and 6) and then penetrated in a semi-circular way into the pin (Figs. 7 and 8). It ends again in a multitude of small torsion fractures (Fig. 9). In this case as well as in that of the cracks at the surface, fatigue fractures are the apparent explanation. All fracture areas are oxidized. It is noteworthy that the zig-zag shaped cracks at the pin bearing surface as proved by the fracture path (Fig. 8) and their branching outward (Fig. 4) did not originate on the surface but in a zone located 2 to 3 mm below it. This is probably

Fig. 1. Crank pin with longitudinal crack (arrow). 0.5 x

A

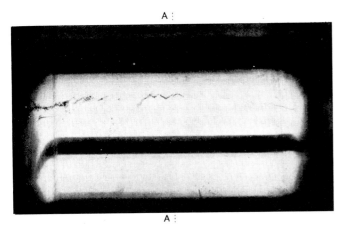

Fig. 2 a. 1,5 x

A

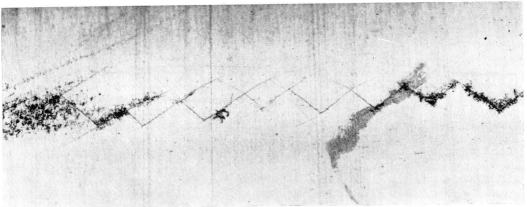

Fig. 2 b. 10 x

Figs. 2 a and b. Longitudinal crack in crank pin

Fig. 3. Section A — A in Fig. 2 a, etched with Nital and after magnetic particle inspection. 1 x

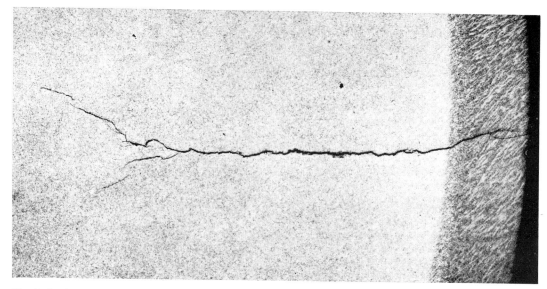

Fig. 4. Crack propagation in transverse section A – – A according to Fig. 2 a, etch: Nital. 8 x

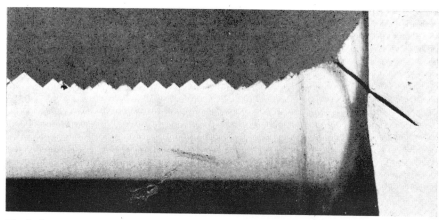

Fig. 5. 3 x

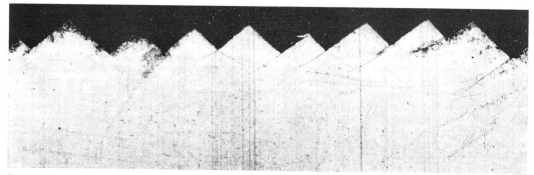

Fig. 6. 10 x

Figs. 5 and 6. Fracture edge of opened crack as seen from bearing surface

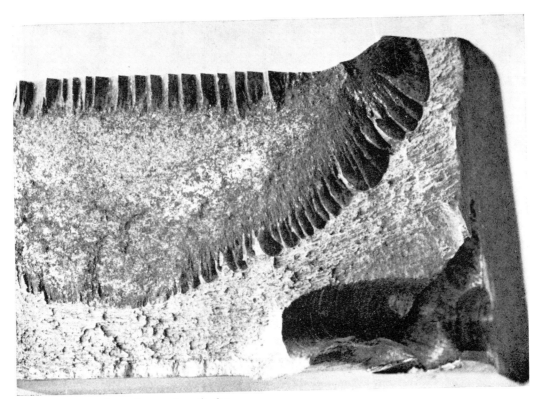

Fig. 7. Fracture surface of opened crack. 3 x

Fig. 8. Torsion fatigue fractures under bearing surface (fracture origin designated by arrows). 10 x

the location at which the sum of the torsional stress (+) and residual compressive stress (−) first exceeded the fatigue strength.

Experience has shown that torsion vibration fractures of this kind usually appear in com- paratively short journal pins at high stresses. This crankshaft fracture presents an example of the damage that is caused or promoted neither by material nor heat treatment mistakes nor by defects of design or machin- ing, but solely by overstressing.

(Continued on the next page)

Fig. 9. Torsion fatigue fractures in fracture end. 10 x

Broken Rim of a
Rolling Mill Transmission

Friedrich Karl Naumann and Ferdinand Spies

Max-Planck-Institut für Eisenforschung

Düsseldorf

The rim of a gear wheel of 420 mm width and 3100 mm in diameter broke after four years of operation time in a sheet bar three high rolling mill. The rim was forged from steel with about 0.4 % C, 0.8 % Si and 1.1 % Mn, oil quenched and tempered to a strength of 65 to 75 kg/mm².

The fracture face is shown in Figs. 1 and 2. The rim started to break in the tooth bottom from a fatigue fracture which extended from the gear side to more than half the rim width. A second incipient failure commenced from the opposite tooth bottom. Both fractures have joined below the tooth of the rim. The failure on the drive side, bottom in Fig. 1 and left in Fig. 2, should have arisen primarily, while the other one is situated at a place normally not tensile stressed and can thus only have generated secondarily due to changed load conditions after the formation of the first incipient crack. The teeth showed many pits on the gear side where the fractures are located, too. However, the pittings were found only on the driven flanks; this indicates that the surface pressing on this side has been larger than on the side of the rolling mill.

Both incipient cracks are, as Figs. 3 and 4 show, fatigue fractures with several starting points which are all located in the transition between tooth flank and tooth bottom. The remaining failure is a fine-grained ductile fracture. The tooth bottom was not only shaped in a sharp-edged way but also roughly machined, too, as is proved by looking at the tooth bottom (Fig. 5) and the cross section through a tooth (Fig. 6). Both faults caused a local rise of tension.

By chemical analysis and by tensile tests of axial and tangential specimens it was observed that the rim had the prescripted composition and strength; only the yield point with 42 kg/mm² did not quite reach the nominal value of 45 kg/mm². Material faults could be found neither macroscopically nor microscopically. The microstructure was a fine-grained batch of pearlite and ferrite, so not really a heat treated structure.

The cross section through a tooth flank on the drive side (Fig. 8) shows a crack which indicates the beginning formation of pittings. These are to be regarded as a type of fatigue fractures which arise by surface pressing due to the alternating shear stresses in cooperation with the lubricant [1]).

Thus the following defects may have contributed to the failure of the rim:

1. The teeth have not supported uniformly over the whole width so that they have been overloaded on one side.

2. The transition from the tooth flanks to the tooth bottom was sharp-edged causing a tension peak there.

3. The tooth bottom was machined only roughly having the same effect as the construction fault of 2.

4. The yield point was a little bit too low. However, since by experience an unsuitable design and rough machining make an increase

1) G. NIEMANN, Maschinenelemente, 2. Bd. Getriebe, Springer-Verlag, Berlin/Göttingen/Heidelberg (1960)

◄
Fig. 1. Location of fracture of the rim with single-sided pitting, gear side on the right. 0.3 x

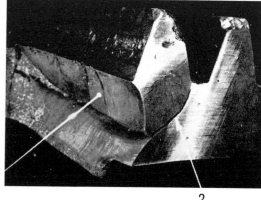

Fig. 2. Fracture face, seen from the gear side. 0.3 x
1 = primary fracture, 2 = secondary fracture

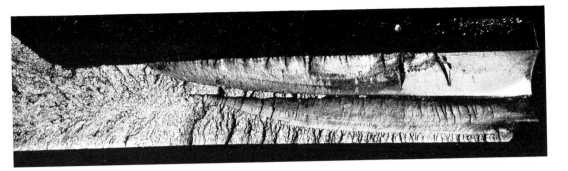

Fig. 3. Primary fracture (1 in Figs. 1 and 2)

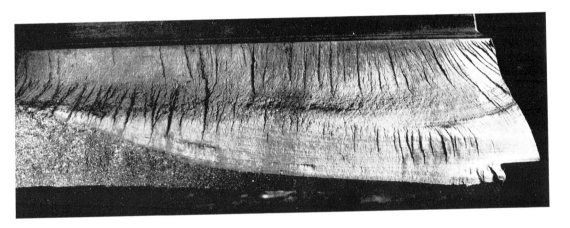

Fig. 4. Secondary fracture (2 in Figs. 1 and 2)

Figs. 3 and 4. Fracture surfaces of the rim; on the right: gear side, on the left: rolling mill side. 0.3 x

Fig. 5. Grooves on the tooth bottom. 2 x

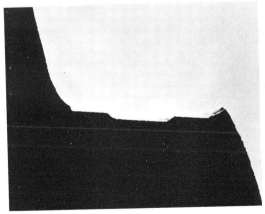

Fig. 6. Profile of bottom of the tooth located next to the fracture. 1 x

Fig. 7. Cross section of the tooth located next to the fracture, sulphur print after Baumann (slightly diminuished)

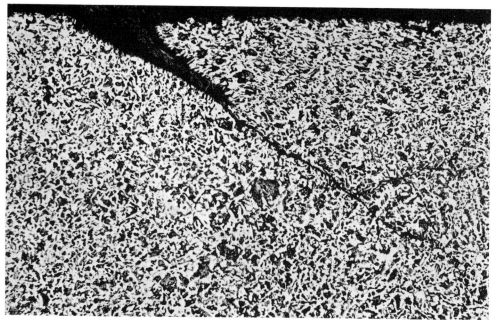

Fig. 8. Cross section of a tooth with pitting, etched with nital. 100 x

in strength rather senseless, that should have been unimportant for the formation of the fractures.

Thus local overload together with poor shaping and machining is to be considered the cause of the failure of the rim.

Fatigue Fractures

Friedrich Karl Naumann and Ferdinand Spies
Max-Planck-Institut für Eisenforschung
Düsseldorf

Vibrational fractures are cracks formed during often repeated, alternating loading at nominal stresses which may be considerably lower than the yield point of the material. The fracture is preceeded by local gliding and the formation of cracks on lattice planes which are favourably orientated with respect to the principal stress (Fig. 1) [1] to [3]). This non-reversible process is often misleadingly called "fatigue". On the whole, a fatigue fracture has the appearance of a brittle fracture. It generally proceeds in stages, which may be large or small according to the value of the stress, and the "arrest lines" so produced, which are arranged concentrically arround the origin of the fracture, give the fracture surface a characteristic appearance. The progress and extent of a vibrational fracture vary according to the nature and degree of the loading [4] [5]). In the case of alternating tension and compression or of bending stresses, the fracture propagates in a plane perpendicular to the principal stress (Figs. 2 to 4) and for torsional stresses at an angle of 45 (Fig. 5). Reverse bending fractures have two origins, or two groups of origins, lying opposite to one another in the cross-section (Figs. 2, 3 and 8); rotational bending fractures usually show several points of origin, distributed round the circumference (Fig. 4). Often several origins lying close together join forces in a step later (Fig. 6). The residual fracture is grainy or fibrous and represents only a small fraction of the fracture surface if the stress was comparatively low (Figs. 2 to 4), but a large fraction if the stress was high (Figs. 6 and 14 a). A vibrational fracture which propagates

Fig. 1. Deformation by slip on the surface of a polished rod specimen (St 37) subjected to reverse bending, with beginning of crack formation (slightly etched during electropolishing). 200 x

Fig. 2. Axle journal with advanced stage of reverse bending fracture. ¹/₃ x

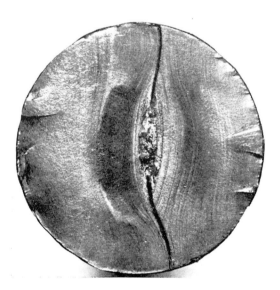

Fig. 3. Fracture of the axle journal of Fig. 2. 1 x

Fig. 4. Rotational bending fracture with small excentric residual fracture. 1 x

quickly can loose its characteristic form and is then no longer recognisable as such with certainty. This is referred to as a "fracture for finite life".

The resistance to vibrational fracture is considerably reduced by "notches", the more so the sharper the notch [6]). Stress concentrations amounting to as much as two or three times the nominal stress can form at notches. In this sense, abrupt changes in cross-section, sharp edges, screw threads, grooves or borings, working faults such as turning or grinding grooves, surface damage such as rubbing or pressure points and stress cracks from grinding or welding [7]) can act as notches. Most vibrational fractures start from such sites of stress concen-

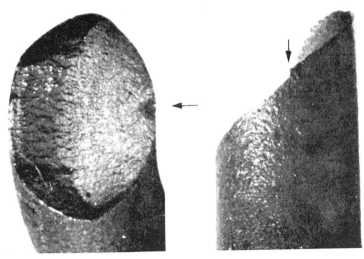

Fig. 5 a. View of fracture surface   Fig. 5 b. Side view

Fig. 5 a and b. Torsional vibration fracture of a valve spring, origin of fracture marked by arrow. 10 x

Fig. 6. Broken cam of a rolling mill clutch with many vibrational fractures originating from a key-groove. Approx. $^1/_2$ x

tration. The higher the strength of a steel, the higher is its "notch sensitivity" (Fig. 7). The advantage of a high strength can be neutralized, or turned to a disadvantage, by inappropriate design or bad working.

Examples of vibrational fractures, the occurrence of which was favoured by the effect of notches, are shown in the following Figs. The cabin bolt in Fig. 8 has broken at a sharp transition in the cross-section. Figure 9 shows an example of the common occurrence of a screw breaking in the first supporting thread. The crankshaft fracture shown in Fig. 10 has started from the hollow groove between the pivot and the cheek; the shaft

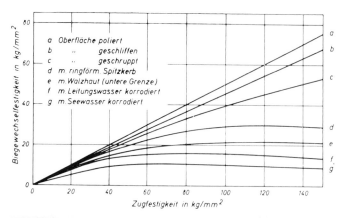

Fig. 7. Relation between the resistance to vibrational fracture and the tensile strength (values taken from [6])

Fig. 8. Broken cabin bolt from an elevator. 1 x

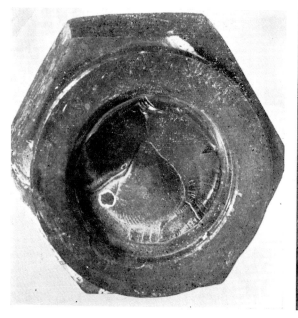

Fig. 9 a. Fracture. 1 x

Fig. 9 b. Longitudinal section, etched in copper ammonium chloride. 2 x

Figs. 9 a and b. Vibrational fracture of a screw in the first supporting thread

Fig. 10. Crankshaft of a compressor, broken in the hollow groove between pivot and cheek. 1 x

Fig. 11. Bending vibrational fracture which had started from a sharp key-groove. 1 x

Fig. 12 a. Fracture with advanced stage of vibrational fracture. ½ x

Fig. 12 b. Sharp transition in cross-section with turning grooves at the origin of the fracture. 1 x

Figs. 12 a and b. Broken rear axle

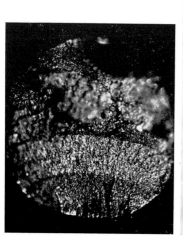

Fig. 13 a. Fracture surface. 20 x

Fig. 13 b. Pressure point at the origin of the fracture. 10 x

Figs. 13 a and b. Broken wheel spoke

Fig. 14 a. Fracture surface. $^1/_2$ x

a

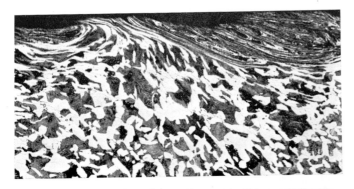

Fig. 14 b. Edge zone near the origin of the vibrational fracture, etched in nital. 80 x

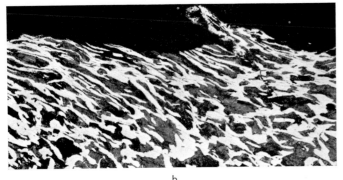

b

Figs. 14 a and b. Broken piston rod of a steam hammer

fracture in Fig. 11 started from the sharp-edged key-groove. The rear axle fracture in Fig. 12 gives an example for the effect of coarse turning grooves in favouring a fracture. Leaf springs frequently break from points which rub against the binding. The fracture of the wheel spokes of a racing car, shown in Fig. 13, were caused by pressure points in the bend below the head. The fracture of the piston rod of Fig. 14 is also due to rubbing. As the microstructure of the fracture shows, the surface layer is not only considerably deformed by cold working, but also shows multiple cracks.

Corrosion scars are notches of a special kind. The combination of vibrational loading with corrosion, even in the case of high-strength steels, leads to fracture at stresses little more than 10 kg/mm^2 (see Fig. 7).

The resistance to vibrational fracture is also lowered by surface decarburization. This effect will be discussed separately in a later paper. The reverse is also true: the resistance can be increased by surface hardening. Since the critical zone in which the stress exceeds the strength, i. e. from where the fracture originates, usually lies just under the surface, the sensitivity to external notches is also reduced in this case.

If it is intended to investigate the cause of vibrational fracture in a given case, the first step, as in all cases of fracture, is to seek the point of origin of the breakage, which is usually not difficult to find. If the vibrational, or other crack has not yet lead to fracture, it should be broken open if possible, since the nature and origin of the crack can often be recognized only in the fracture surface. If the crack has started from outside, the next step is to inspect the surface, and if it

comes from a transition in the cross-section, the constructional form of this transition is also noted. If the latter is sharp-edged, or if the surface displays coarse working grooves at the point of origin of the fracture, rubbing points or other damage, it is possible to state that these features have at least favoured the fracture. (The judgement "caused" must usually be avoided by the metallographer, since the actual loading which led to the fracture is known only in rare cases). If a surface decarburization can be seen in a polished surface through the fracture, the same applies to this. The search for a material fault is almost always fruitless in such cases. The situation is different if the fracture has started from a point within the material. Then material faults such as contraction cavities, marked segregation or flakes are often found at the origin of the fracture. Such cases have been reported previously [8], and will also be treated in future papers.

[1] M. HEMPEL, E. HOUDREMONT; Beitrag zur Kenntnis der Vorgänge bei der Dauerbeanspruchung von Werkstoffen, Stahl und Eisen 73 (1953) 1503/11

[2] M. HEMPEL, Verformungserscheinungen und Rißbildung an biegewechselbeanspruchten Rundproben eines vergüteten Chrom-Molybdän-Stahles, Arch. Eisenhüttenwes. 37 (1966) 887/95

[3] M. HEMPEL, Die Entstehung von Mikrorissen in metallischen Werkstoffen unter Wechselbeanspruchung, Arch. Eisenhüttenwes. 38 (1967) 446/55

[4] M. EHRT, G. KÜHNELT, Das Gesicht des Dauerbruches. Mitt. d. Materialprüfst. d. Allianz u. Stuttgarter Verein Vers. A.G., Heft 4, Berlin (1938)

[5] G. RICHTER, Was sagt die Ausbildung eines Bruches über die Bruchursache aus? Der Maschinenschaden 29 (1956) 97/106

[6] Werkstoffhandbuch Stahl und Eisen, 4 Aufl., Düsseldorf (1965) Blatt D 11–7

[7] F. K. NAUMANN, F. SPIES, Prakt. Metallogr. 5 (1968) 294

[8] F. K. NAUMANN, F. SPIES, Prakt. Metallogr. 4 (1967) 541/46

Oxidized Recuperator Pipes

Friedrich Karl Naumann and Ferdinand Spies

Max-Planck-Institut für Eisenforschung

Düsseldorf

Pipes 42.25 x 3.25 mm from a blast furnace gas fired recuperator for the preheating of air were heavily oxidized and perforated in places. Depending on the load the temperature of the combustion gas in the flue channel on entering Group 1 of the recuperator was 1150 to 1200° C, falling in Group 1 to 1125° C, in Group 2 to 1050 to 1000° C and in Group 3 to 860 to 840° C. The air was heated to 420 to 450° C in Group 3, to 580 to 640° C in Group 1 and to 750 to 820° C in Group 2. Theoretically the maximum wall temperature in Group 1 should be 790 to 920° C, in Group 2 860 to 960° C and in Group 3 680 to 770° C. In practice a maximum temperature of 1025° C was measured. The pipes in Groups 1 and 2 should have been made from a steel with ⩽ 0.18% C and 27 % Cr, those in Group 3 from a steel with ⩽ 0.12 % Cr, 6 to 7 % Cr and 0.5 to 1.0 % Al such as X10 Cr Al 7, Material No. 1.4713.

Two pipes were examined from each group. Pipes from Groups 1 and 2 were heavily oxidized on the gas side (externally) and had burnt through in places (Fig. 1). The pipes from Group 3 however had only a thin external scale layer. Internally, on the air side, all the pipes were smooth.

Chemical analysis of the pipes yielded the data shown in Table 1. In the case of the pipes in Groups 1 and 2 the carbon content is much higher than specified and in the case of the pipes in Group 3, the aluminium content lower.

Baumann prints were taken of the external surface and of cross sections of the pipes to test for sulphur present as sulphide (Figs. 2

Fig. 1. View of recuperator pipes from Groups 1, 2 and 3 (top to bottom). 1/2 ×

Table 1. Chemical composition of the pipes

pipe from group	C %	Si %	Mn %	Cr %	Al %
1	0,37	0,29	0,50	26,1	0,05
2	0,29	0,45	0,63	25,9	0,01
3	0,12	0,82	0,44	6,58	0,11

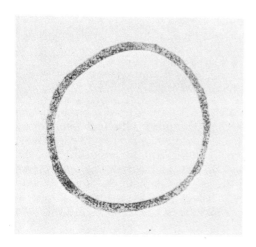

Fig. 2. Pipe from Group 1

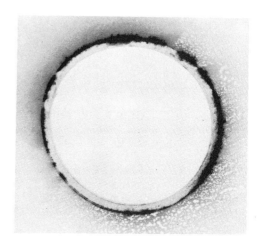

Fig. 3. Pipe from Group 3

Figs. 2 and 3. Baumann sulphur prints from cross sections

and 3). A reaction was obtained only with pipes from Group 3.

Microscopic examination revealed a thick oxide layer on the external surfaces of pipes from Groups 1 and 2 below which there was a narrow zone with oxide and sulphide precipitates (Fig. 4). The internal surface, however, was coated with only a thin layer of scale (Fig. 5). Pipes from Group 3 had an external oxide coating below which there was a layer of sulphide scale which had preferentially penetrated the grain boundaries (Fig. 6). On the internal surface (air side) they had a coating of oxide scale (Fig.7). Like the analysis, the microstructure of the pipes from Groups 1 and 2 showed a higher than specified carbon content of the Group 3 pipes, howe-

ver, was scarcely more than normal for this steel (Fig. 9).

It must be concluded from the findings that the blast furnace gas had a high sulphur content. Both the carburization and the formation of sulphide prove that in addition, from time to time at least, combustion was incomplete and the operation was carried out in a reducing atmosphere, with the result that oxygen deficiency prevented the formation or maintenance of a protective surface layer on the external surface of the pipes. The sulphur would probably not have damaged the nickel-free steel used here at the given temperatures if it had been present as sulphur dioxide in an oxidizing atmosphere. The damage was therefore caused primarily by an incorrectly conducted combustion process.

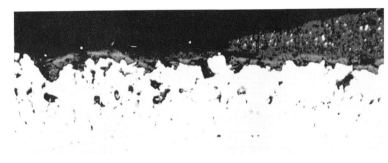

Fig. 4. External surface (gas side)

Fig. 5. Internal surface (air side)

Figs. 4 and 5. Surface microstructure of a pipe from Group 2.
Cross section, unetched. 200 × .

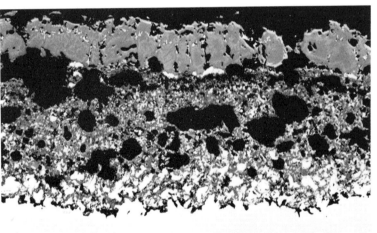

Fig. 6. External surface (gas side).

Fig. 7. Internal surface (air side)

Figs. 6 and 7. Surface microstructure of a pipe from Group 3.
Cross section, unetched. 200 × .

(Continued on the next page)

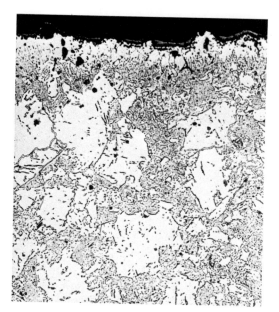

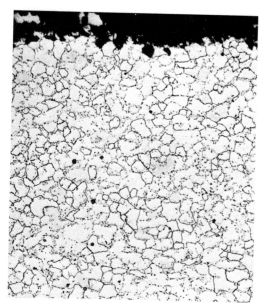

Fig. 8. External surface microstructure of a pipe from Group 2.
Cross section, etched in V2A pickle. 200 ×

Fig. 9. External surface microstructure of a pipe from Group 3.
Cross section, etched in V2A pickle. 200 × .

Damaged Copper Hot Water Pipe

Egon Kauczor

Staatliches Materialprüfungsamt an der Fachhochschule
Hamburg

A general view of the specimen under examination from a copper hot water system is shown in Fig. 1 in the as supplied condition. As can be seen from the photograph, a bent pipe has been soldered into a straight pipe with twice the diameter. The neighbourhood of the soldered joint was covered with corrosion product predominantly blue-green in colour, presumably carbonates. When these corrosion products were scratched off it was seen that the copper beneath this layer had not suffered noticeable attack.

The object of the examination were the localised deep cavities located almost symmetrically to both sides of the inserted end of the narrower tube on the internal wall of

the wider tube which had in one place been eaten right through. The water seeping out at this point could be the cause of the external deposits.

Figure 2 shows one of the cavities revealed by a longitudinal section of the straight pipe. An unetched polished macrosection taken at ← is shown in Fig. 3. A piece of the soldered-in pipe and the soldered joint can be seen on the left and the wall of the wider tube which had been eaten right through on the right.

The symmetrical location on each side of the point of insertion of the narrower pipe and the localised sharp delineation of the attack indicate erosion due to the

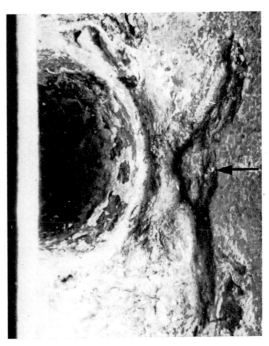

Fig. 1. General view of the pipe joint under examination in the as supplied condition. ¹/₂ x

Fig. 2. Removal of material from the internal wall of the wider pipe close to the inserted end of the narrower pipe. 3 x

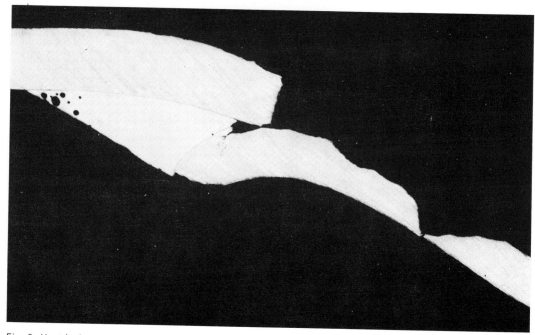

Fig. 3. Unetched polished section taken through the point marked ← in Fig. 2. 20 x

formation of turbulence. Turbulence can cause localised rupture and removal of the natural protective layer on the copper. A potential difference is set up between the exposed metal surface and its surroundings still covered by the protective layer. The anodised unprotected region is thereby subjected to corrosive influences.

The joint action of mechanical wear due to turbulence and simultaneous corrosive attack can lead to the removal of such large thicknesses of material. This process can be accelerated by finely dispersed solids carried along in the water and also by dissolved gases which form bubbles at the transition from the small to the large cross section.

Such phenomena can be combated most effectively if by avoiding sharp transitions and abrupt changes in cross section it is possible to design the pipe work so that localised turbulence is obviated. Degassing and cleansing of the water also reduce the

danger of erosion particularly in the case of softened water which takes up oxygen and carbon dioxide very readily thus becoming particularly aggressive.

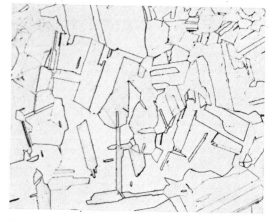

Fig. 4. Microstructure of a polished specimen from the wide pipe etched with ammonium persulphate. 200 x

Examination of a Blistered and Cracked Natural Gas Line

Friedrich Karl Naumann and Ferdinand Spies

Max-Planck-Institut für Eisenforschung

Düsseldorf

A welded natural gas line of 400 mm O. D. and 9 mm wall thickness made of unalloyed steel with 0.22 % C had to be removed from service after 4 months because of pipe burst whose cause was unknown. Examination of the pipe showed damage which had no connection with the fracture of the line. In order to investigate these damages, test sections were cut at points 3, 6, and 10 km distant from the entrance of the 12 km long pipe line that was operated under 60 atm. pressure.

Where the pipes had not rusted their inner surfaces showed a black film, and also large flat blisters (Fig. 1) when observed obliquely. Ultrasonic testing also showed separations at many places where no blisters could be discerned visually. Only the butt welds and their adjoining areas proved to be crack-free. The black film developed large quantities of hydrogen sulfide when drops of sulfuric acid were applied and dyed bromide paper saturated with sulfuric acid (Baumann Sulfur Print) dark (Fig. 2). Therefore it consisted of iron sulfide. The blisters gave off gas under considerable pressure when drilled open under acidic water. The gas contained 89.4 % H_2, 2.4 % CO_2, 7.0 % N_2 and 1.7 % O_2. If the nitrogen-oxygen mixture that probably consists of air which penetrated during the capture of the gas, is disregarded, the original gas contained 97.4 % H_2.

Cuts were made for the metallographic examination from all sections, preferably at those locations where separations existed according to ultrasonic testing.

Fig. 1. Inner view of section 1 (3 km from gas entrance). Pipe diameter 400 mm

The pipe section located next to the gas entrance was permeated by cracks or blisters almost over its entire perimeter in agreement with the ultrasonic test results (Fig. 3.). Only the weld seam and a strip on each side of it were crack-free. The separations were located mostly in the center of the pipe wall, and ran principally in the directions of the fiber, and approached the inner surface in a stepwise manner propagating from fiber to fiber. The sections 2 and 3 at 6 and 10 km distance from the gas entrance showed fewer cracks of this nature. Their position again coincided with the results obtained with the ultrasonic tester.

All sections were also investigated microscopically, both transversely and longitudi-nally. They contained long drawn-out, thin sulfidic and oxidic inclusions as is customary with sheet and strip. The cracks propagated from these and then followed along the stress lines jumping from one slag streak to the next (Figs. 4 to 6).

According to this investigation, the natural gas contained hydrogen sulfide. This attacks steel when it occurs in aqueous solution according to an electrochemical process described by F. K. Naumann and W. Carius[1]) and forms iron sulfide and hydrogen. The hydrogen penetrates the steel and precipitates molecularly at weak points such as inclusions, grain- and phase boundaries at a pressure corresponding to its concentration and tears the metal apart. In soft

Fig. 2. Sulfur print according to Baumann of the inner surface of the line

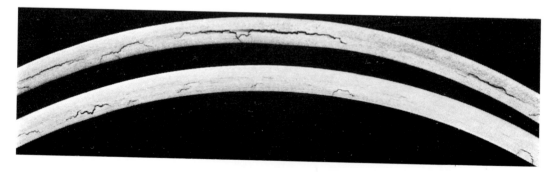

Fig. 3. Cross section of section 1, etch: copper ammonium chloride (according to Heyn). approx. 0.7 x

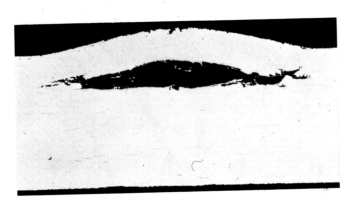

Fig. 4. Small blister. 25 x

Fig. 5. Fracture along slag streak. 100 x

Fig. 6. Crack propagation from slag streak to streak

Figs. 4 to 6. Longitudinal section, unetched

Fig. 7. Fracture location of the pipe (pipe diameter 400 mm), fissured areas marked by chalk after ultrasonic testing

Fig. 8. Inner surface next to fracture (top). 1 x

Fig. 9. Fracture area. 1 x

Figs. 7 to 9. Burst test

steels of large strain capacity as are discussed here, separations and blisters are formed far below the yield point of the steel and an additional tensile strain may lead to fractures as well[2] [3]). High strength steels with correspondingly low elongation do not form blisters but fracture right away[4]). This process is comparable therefore to that of blisters- and crack formation during pickling[5]).

In order to ascertain how high the mechanical stresses can be that a gas line under the above described conditions could withstand, a 4 m long section was tested of the second kilometer of the line. It was closed off at both ends and put under water pressure. Prior to this it was tested ultrasonically for interior defects and proved to be extremely defective. The pressure could be raised without interruption to 90 at. initially, e. i. 1½ times the operating pressure, without fracture initiation or observable elongation. Subsequently the water pressure was increased in steps of 5 at. Only at 155 at. corresponding to a tangential tensile stress of 34 kp/mm², did the inner pressure drop significantly to 150 at. This value corresponds almost exactly to the transverse yield point of 33.8 kp/mm² of the sheets used, that was determined by averaging 47 measured values. Before the tensile strength of the sheets of 49.2 kp/mm² was reached, the pipe burst open prematurely after expanding locally, at 170 at. = 37.6 kp/mm² tangential stress, i. e. soon after passing the yield point.

Figure 7 shows the fracture zone. The frac- ture is located in a plane approximately 120° from the weld seam. The area had proved to be extremely defective according to prior ultrasonic testing, and it was therefore marked with chalk and shaded. Next to the fracture and parallel to it a number of further cracks were formed or opened up (Fig. 8). The fracture (Fig. 9) is fissured and fibrous, but it still shows clearly signs of deformation that occurred prior to crack formation.

Based upon this investigation the pipe line was taken out of service and reconstructed. In order to avoid a repetition of this failure, the following two measures offer possibilities, namely firstly to desulfurize the gas. Tests showed however, that the desulfurization would have to be carried very far in order to be assuredly successful[6]). The second possibility consisted in drying the gas to such a degree that no formation of a condensate at the lowest expected winter temperature could occur, since the corrosion process, being of electrolytic nature, is dependent upon the presence of a liquid phase. This measure was deemed more effective and cheaper, and was recommended.

Literatur/References

1) F. K. NAUMANN, W. CARIUS, Archiv Eisenhütten- wes. 30 (1959) 233/238; 383/392; 361/370
2) W. DAHL, H. STOFFELS, H. HENGSTENBERG, C. DÜREN, Stahl und Eisen 87 (1967) 125/136
3) H. SCHENK, E. SCHMIDTMANN, H. F. KLÄRNER, Stahl und Eisen 87 (1967) 136/146
4) F. K. NAUMANN, Erdöl-Zeitschr. 73 (1957) 4/14
5) Prakt. Metallographie 4 (1967) 663/670
6) F. K. NAUMANN, Stahl u. Eisen 87 (1967) 146/154

Damaged Cylinder Lining from a Diesel Motor

Egon Kauczor
Staatliches Materialprüfungsamt an der Fachhochschule
Hamburg

The cylinder lining under examination had suffered localised damage on the cooling water side leading to serration of the edges and heavy pitting as shown in Fig. 1. Beyond this heavily damaged zone, the external wall was coated with deposits shown by qualitative chemical analysis to consist principally of Fe_3O_4 (magnetite) with small amounts of SiO_2, graphite and CaO. The material beneath the deposit had not been attacked, which suggests that foreign particles were responsible. These were possibly carried round by the circulating cooling water but could also have originated in part from the damaged zone as indicated by the presence of graphite and SiO_2.

Fig. 1. General view of cylinder lining as supplied. 1/3 x

Fig. 2. Microstructure of specimen taken from cylinder lining. Etched in 2 % nital. 500 x ▶

It was established metallographically that the cylinder lining was made of cast iron. The microstructure is reproduced in Fig. 2.

This heavy damage is cavitation damage, frequently observed in diesel motor cylinders.

Cavitation must always be reckoned with on a body vibrating in a liquid when the amplitude and frequency attain such high values that the sluggish liquid can no longer keep pace. As the body returns, vacuum bubbles form which implode on the next outward movement of the medium. This produces such a high pressure on a microscopically small area that the material disintegrates and finally particles of material are knocked out of the surface.

The conditions for cavitation are set up at the cylinder linings in a running diesel motor because at the moment of reversal the cylinder is excited to bending vibrations by the sideways pressure of the piston. If the cooling water can no longer follow the vibration of the cylinder vacuum bubbles form at the most strongly vibrating point (impact side) which implode on reversal of the vibration direction and produce cavities on the

cylinder surface as a result of this violent cyclic stressing.

In order to combat such damage the following measures are recommended in the specialist literature

1) reduction in piston play

2) reduction in the amplitude by thicker walled linings

3) hard chromizing of the cooling water side

4) addition of a protective oil to the cooling water

The effect of the protective oil is presumably based on a film of oil which forms on the cylinder surface and which is not so easily scoured off during vibration. The effect of the imploding vacuum bubbles is reduced by the oil film which can renew itself from the emulsion.

Literatur/References

1) E. WITZKY, Erfahrungen mit Dieselzylinder-Kavitation-Pitting in USA, MTZ Motortechnische Zeitschrift 18 (1957) Nr. 2, 33/37

2) H. BULLNHEIMER, Kühlwasserseitige Anfressungen an Zylinderbüchsen und Gestellen schnellaufender Dieselmotoren, MAN-Forschungsheft Nr. 9 (1960)

Leaky Derusting Vessel Made of 18/8 Steel

Friedrich Karl Naumann and Ferdinand Spies

Max-Planck-Institut für Eisenforschung
Düsseldorf

A welded vessel made of acid resistant steel of the 18/8 type used in the derusting of motor vehicle parts started to leak after a long period due to the formation of cracks. The vessel was heated from the outside and so that it was not in direct contact with the flame it was surrounded by a casing of unalloyed steel.

Figures 1 and 2 show one of the cracks from the outside and from the inside. In addition to the penetrated crack, a multitude of small cracks are visible on the outer surface. A metallographic section through this region confirms that the cracks propagated from the outside (Fig. 3). The inner surface is completely unscathed. Where

Fig. 1. Outside of vessel. 1 x

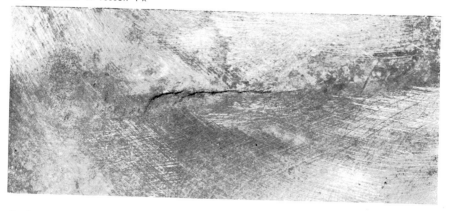

Fig. 2. Inside of vessel. 1 x

Fig. 3. Metallographic section through cracked region, unetched, above: outer surface of vessel. 10 x

Fig. 4. Cracked region on outer surface of vessel, unetched. 100 x

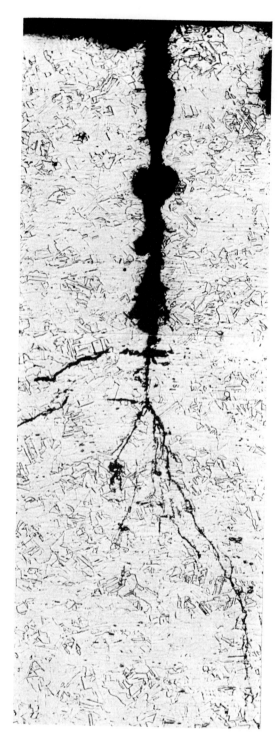

the cracks have not yet been eroded away it is clear that they run transcrystalline (Figs. 4 and 5). This is clearly a case of stress corrosion cracking. Since the cracks propagated from the outer surface of the vessel they cannot have been caused by the derusting agent but only by the external atmosphere in conjunction with welding stresses. This is characteristic of this selective type of corrosion; it actually occurs preferentially as a result of mild corrosive agents.

The narrow gap between vessel and mild steel casing may have aggravated the situation in that it hindered ventilation and evaporation of condensation and favoured the absorption and concentration of acids and salts. Contact and crevice corrosion due to deposition of rust from the mild steel casing could have contributed.

Fig. 5. As Fig. 4., after etching with V2A pickle. 200 x

Scaling of Resistance Heating Elements in a Through-Type Annealing Furnace

Friedrich Karl Naumann and Ferdinand Spies

Max-Planck-Institut für Eisenforschung
Düsseldorf

Heating elements, consisting of strips, 40 mm x 2 mm, of the widely used resistance heating alloy with 80 % Ni and 20 % Cr, and designed to withstand a temperature of 1175 ° C, were rendered unusable by scaling after a few months service in a through-type annealing furnace, although the temperature is supposed not to have exceeded 1050° C. The strips, wound in a zig-zag form, were situated under the material to be annealed. The latter consisted of dynamo and transformer sheets, which were annealed in a decarburising atmosphere of moist hydrogen.

Externally, the strips showed wart-like surface upheavals, lying preferentially near the edges and bends (Fig. 1), and also edge cracks (Fig. 2). In the fracture, internal voids were visible (Fig. 3), which gave the strips a doubled appearance. The cracks were coloured green by chromium oxide, except for a narrow, bright, metallic strip in the core.

In the transverse and longitudinal sections, it was seen that the strips were heavily scaled under the warts (Figs. 4 and 5). The attack had started from isolated points where the protective layer had been punctured, and had then advanced radially from these points. Next to the warts were regions where the surface was completely intact. When the scaling had started from two opposite points on the strip, voids were formed in the middle, which are in some cases wide open.

Fig. 1. Wart-like upheavals. 0.7 x

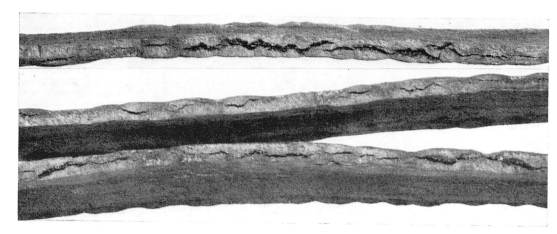

Fig. 2. Edges with cracks. 2 x

Fig. 3. Fractures with internal voids. 1 x

Figs. 1 to 3. Views of the scaled heating element

The microstructure of the scale layer is shown in Fig. 6. The oxidation has apparently proceeded initially on the austenite grain boundaries. The first voids formed on the grain boundaries (Fig. 7). These observations suggest melting as a possible cause of the failure, but no traces of a eutectic structure formed on solidification were detected.

At a certain distance under the scale layer, a hard phase, which clearly showed relief in phase contrast (Fig. 8), had precipitated at the austenite grain boundaries. It was stained dark by Murakami's ferricyanide solution, and also by electrolytic etching in 0.1 N sodium hydroxide [1]), showing that it consisted of a chromium-rich carbide. This carbide is also present in the original structure of the alloy, but in appreciably smaller amounts (Fig. 10).

The structural observations indicate a special case of internal oxidation. By this we understand an internal oxidation in which

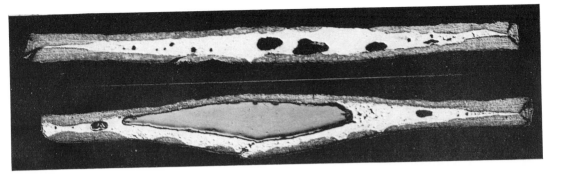

Fig. 4. Cross-section, unetched. 3 x

Fig. 5. Longitudinal section, unetched. 3 x

not all the components of the alloy are involved, so that some of the metal remains unchanged. The required conditions for this are apparently provided by the moist hydrogen atmosphere of the annealing furnace, in which the chromium is oxidised, whilst the oxides of iron and nickel are reduced. Even the carbon suffers incomplete combustion, and is enriched in the core. Thus no protective layer can form or be maintained, and at places where the original layer was damaged, it cannot form again. The preferential distribution of the breakages at the edges and bends of the strips indicate that the protective layer which was initially formed during the manufacture of the strips

had been ruptured locally at first by mechanical or thermal stresses.

The intergranular advancement of the oxidation may have been favoured by the precipitation of chromium-rich carbides on the austenite grain boundaries. The oxidised chromium must be replaced by diffusion to the reaction front. This leaves behind vacancies, which can agglomerate to form larger voids. The latter can then be blown up to the size seen in Fig. 4 by the penetration of atomic hydrogen from the dissociation of water vapour and subsequent molecular precipitation under high pressure.

Fig. 6. Structure of scale layer, edge-core-edge, cross-section, unetched. 100 x

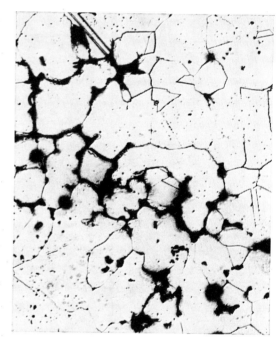

Fig. 7. Voids on the austenite grain boundaries below the scaled region, cross-section, etched in V2A pickling solution. 100 x

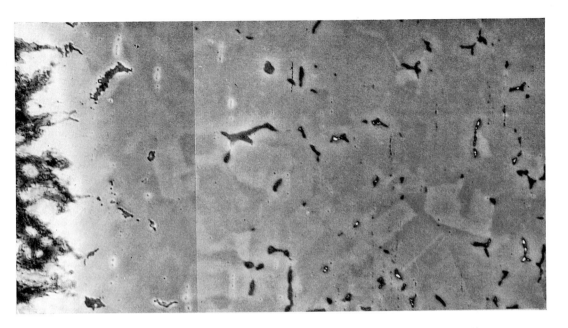

Fig. 8. Precipitates under the scale layer, longitudinal section, unetched, phase contrast micrograph (−1). 100 x

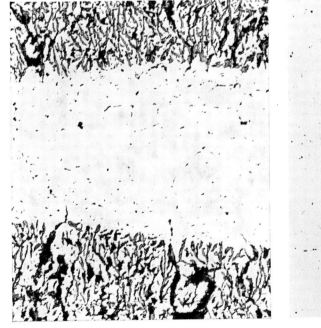

Fig. 9. Precipitates under the scale layer, cross-section, etched in ferricyanide solution (after Murakami). 100 x

Fig. 10. Precipitates in the original structure of the heating strip, cross-section, etched in ferricyanide solution (after Murakami). 100 x

This form of internal oxidation is, in the case of Ni-Cr alloys, known as green rot. Alloys containing iron should be more resistant. As a preventive measure it was recommended to reduce the operating temperature of the strip sufficiently to allow the use of Fe-Ni-Cr alloys. Alternatively, when the strip is taken into service it should be heated to a high temperature in air, before introducing the hydrogen atmosphere, and oxidised for some time; this treatment should be repeated later from time to time.

1) F. K. NAUMANN, Arch. Eisenhüttenwes. 38 (1967) 463/468

Examination of an Oxidized Heating Coil

Friedrich Karl Naumann and Ferdinand Spies

Max-Planck-Institut für Eisenforschung
Düsseldorf

A coil made of the nickel-chromium alloy NiCr 80/20 (Material No. 2.4869) with approx. 80 % Ni and 20 % Cr had burnt through after a brief period of operation as heating element in a brazing furnace. The protective atmosphere consisted of an incompletely combusted coal gas. Furnace temperature reached 1150°C.

The spiral was covered by a dark green chromium oxide layer that had penetrated deeply into the wire in spots as could be seen from the fractures.

At a slightly oxidized spot, chips were taken for an analysis of the wire composition after the oxide layer was removed from the surface as far as possible. Quantitative analysis showed the following (wt.%):

Cr	Ni
13,7	82,2
Mn	Fe
0,05	0,5

Chromium contents was higher while nickel contents was lower than standard for this alloy. This can be explained by preferential combustion of chromium and its effusion to the place of reaction.

The oxide contained 0.011 % S and 0.009 % N. This ruled out a sulphur-containing gas component as cause for the fast and strong oxidation.

Metallographic examination confirmed that the attack emanated from the individual centers and penetrated deeply into the material locally (Fig. 1). It progressed preferentially

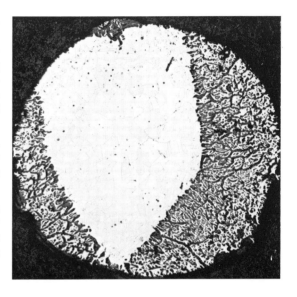

Fig. 1. Transverse section through a strongly oxidized spot of the spiral. Etch: V2A pickle. 15 ×

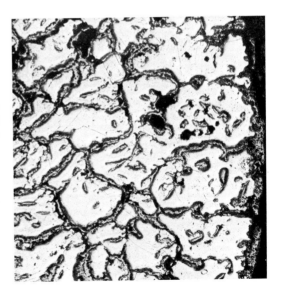

Fig. 2. Peripheral zone from Fig. 1. 75 ×

towards the austenitic grain boundaries (Fig. 2). The oxide had the typical green color of chromium oxide. The wire was permeated by pores up to the middle of the cross section (Figs. 3 and 4) comparable to a previously described defect[1]). The pores may probably be interpreted as vacancy concentrations as a consequence of effusion of the chromium. The austenite was very coarse-grained due to overheating.

This type of selective oxidation at which the easily oxidized chromium burns, while the nickel is not attacked, is caused by mildly oxidizing gases and is sometimes designated as green rot. Under these conditions chromium-containing steels and alloys whose oxidation resistance is based upon formation of tight oxide layers are not stable.

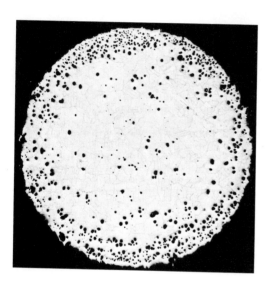
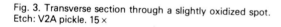

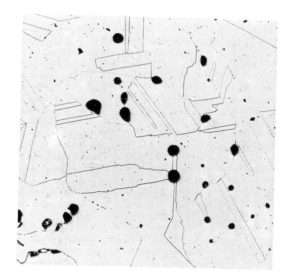

Fig. 3. Transverse section through a slightly oxidized spot. Etch: V2A pickle. 15 ×

Fig. 4. Core structure of the spiral. Transverse section. Etch: V2A pickle. 75 ×

Corroded Pipe Section of Oil Burner for Superheated Steam Generator

Friedrich Karl Naumann and Ferdinand Spies

Max-Planck-Institut für Eisenforschung

Düsseldorf

A heat exchanger made of a pipe of 35/30 mm diameter, in which oil was heated from the outside from approximately 90°C to 170°C, by superheated steam of about 8 to 10 atmospheres had developed a leak at the rolled joint of the pipe and pipe bottom. The pipes were supposed to be made from St 35.29 steel and annealed at the rolled joint to 100 mm length. An examination of the leaky pipe end was to determine whether the material used was according to standard specification and whether the annealing and rolling operations were done correctly.

The outer pipe surface was strongly pitted by corrosion all around the rolled joint (Fig. 1). In the vicinity of the steam chamber the pipe wall had oxidized through from the exterior to the interior at one spot (Fig. 2). Adjoining this spot grooves caused by erosion were noticeable.

This is a typical case of crevice corrosion. The rolled joint evidently was not entirely tight, so that saturated steam condensate could penetrate into the gap. In such crevices so-called aeration elements are formed which are caused by oxygen depletion. Their occurrence greatly accelerates corrosion. This may be expected especially in those instances

Figs. 1 and 2. External surface of pipe section, seen from 2 sides. Right: Rolled joint. 1 ×

where the oil is still relatively cold and the steam has cooled off substantially.

The structure of the pipe end consisted of a fine-grained mixture of ferrite and pearlite with approx. 0.15 % C (Fig. 3). The pipe therefore could have been made of the steel ordered. An exact determination of the standard quality for which mechanical testing would have been necessary was not possible with the small pipe end at hand. Mechanical testing could be omitted also because the failure was apparently due to corrosion, irrespective of the material. It was facilitated by not rolling a tight joint.

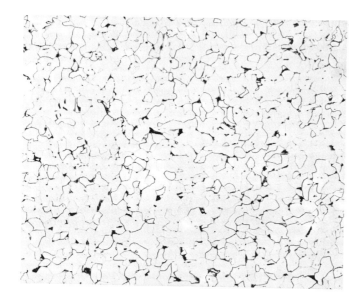

Fig. 3. Structure at the rolled joint. Longitudinal section. Etch: Nital. 100 ×

Decarburisation

Friedrich Karl Naumann and Ferdinand Spies

Max-Planck-Institut für Eisenforschung
Düsseldorf

One reason for many damages is the decarburisation of steel. This may occur in two different forms, namely as skin decarburisation by gases either wet or containing oxygen, and as a deep going destruction of the material by hydrogen under high pressure.

The reaction of the oxygen decarburisation takes place by the formation of carbon oxide on the surface. In slightly oxidizing atmospheres like wet hydrogen the velocity of decarburisation is determined by the diffusion velocity of carbon in iron, however in the presence of more oxidizing gases like oxygen or wet nitrogen the removal of the carbon oxides may be obstructed by the formation of scale and thus decarburisation will be stopped [1]. So, if in view of its consequences a decarburisation should by all means be avoided, it is not advisable to perform a heat treatment in slightly oxidizing atmospheres as they can always generate in annealing boxes or protective gases with small amounts of air or moisture present.

How does a skin decarburisation become evident to a metallographer? It can go so far that the carbon has completely disappeared at the outer edge. After DIN 17 014 this state is called „Auskohlung". This form of a complete decarburisation with a steep transition to the core material is very likely to formate at low temperatures in the α- or $(\alpha+\gamma)$-area of iron (cases a and b in Fig. 1 and Fig. 2 a). When the decarburisation starts in the γ-area and — during the procedure — passes through the $(\alpha+\gamma)$-area (case c in Fig. 1), the totally decarburised zone is followed by a partially decarburised transition zone (Fig. 2 b). The ferrite precipitated while passing through the $(\alpha+\gamma)$-area takes up a columnar structure. This form of abrupt decarburisation is usually not observed when annealing in gases producing scale, since there because of the hindrance by scale decarburisation at such

low temperatures proceeds not at all or very slowly — as already mentioned above. An exception is made by steels with alloying elements, which cut off the γ-area; thus the α- and $(\alpha+\gamma)$-area extend to elevated temperatures. To these steels belong for example the silicon spring steels and the Al alloyed nitriding steels (Figs. 7 and 8). When the whole decarburisation proceeds in the γ-area, that is above 911° C in unalloyed steels (case d in Fig. 1), the carbon content goes over gradually — with or without a zone of unorientated ferrite — into that of core (Fig. 2 c). So it is possible from the structure of the decarburised zone to draw certain conclusions regarding the conditions for its formation.

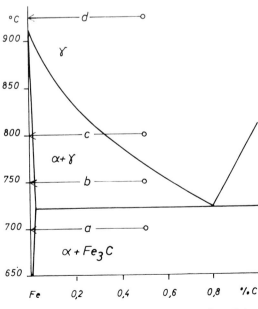

Fig. 1. Possible cases of decarburisation, schematicly

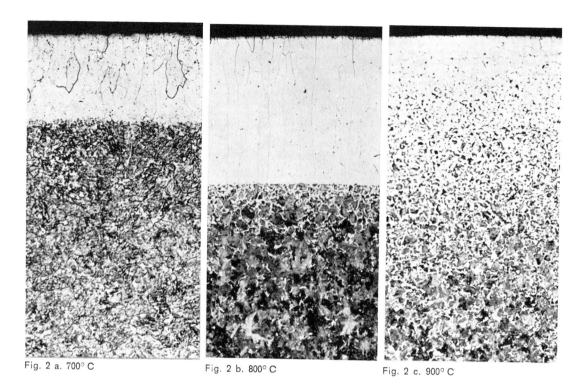

Fig. 2 a. 700° C   Fig. 2 b. 800° C   Fig. 2 c. 900° C

Figs. 2 a to c. Edge structure of a steel with 0,73 % C after 4 h annealing in wet hydrogen of 1 atm pressure, etched in picral. 100 x

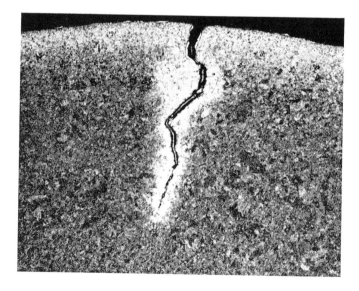

Fig. 3. Longitudinal crack in a hardened and tempered bar of nickel-chromium steel, cross section, etched in nital. 10 x

For the examination of cracks it is important to know when or during which working process the crack has generated. To that, a hint may be given by the structure of the crack edges. If the crack is severely decarburised on the whole length (Fig. 3), it can be surely assumed that it did not generate during the last heat treatment like quenching or tempering but was already existent earlier and formated at the latest during warming up for the

Fig. 4. Longitudinal crack in a hardened and tempered spring washer of silicon steel, cross section, etched in picral. 100 x

probability a hardening crack which started from a surface defect, like a rolling fold for example. But also the opposite case may occur, namely that a crack is scaled but not decarburised in its outer part, however not scaled but clearly decarburised in its inner branch (Fig. 5 a). As mentioned introductorily, here the scale which formated under heavily oxidizing conditions near the surface prohibited the decarburisation while the less oxidizing atmosphere in the narrow crack branch led to a decarburisation. If the decarburisation has proceeded up to the formation of a purely ferrite edge which passes over to the core structure abruptly, like in Fig. 5 b, it may be concluded that it came into existence at relatively low temperatures. Also then the crack can be a hardening crack which has been decarburised during tempering.

How are the properties of work pieces changed by skin decarburisation? A totally decarburised layer has a lower hardness and resistance to wear than the basic material. It is not at all or not so well hardenable (Fig. 6). The different transformation behaviour of edge and core may lead to tensions and as a consequence to distortions or cracks. During nitriding decarburised surfaces tend to peeling off (Fig. 7). The fatigue strength will be decisively decreased by skin decarburisation (Fig. 8). Thus fatigue failures in vibrating machine parts like springs caused by skin decarburisation are not a rarity [2] [3] [4]. However the soft edge layer formated by oxygen or vapour decarburisation is not brittle or cracky.

final treatment. If in a hardened or tempered work piece a broad crack severely decarburised on the outside goes suddenly over into a dense, closed, not decarburised crack with a jagged path (Fig. 4), it is with great

Quite different are procedure and effect of a hydrogen decarburisation. The decarburisation of dry hydrogen is lower by one order of magnitude than of wet hydrogen [5]. Decarburisation via the methane reaction under atmospheric pressure becomes only notable at temperatures beyond 1000° C [1]. It is greatly favoured by an increase of the hydrogen pressure and thus by a rising dissociation of the hydrogen. At pressures from 100 to 1000 atmospheres, as nowadays common in technical hydrogenation processes, failures of

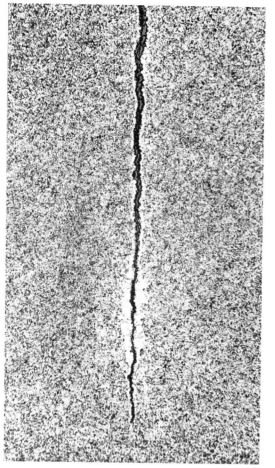 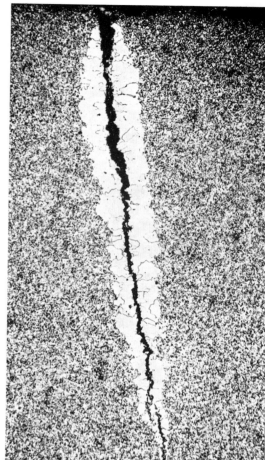

Fig. 5 a. Branch of a long crack

Fig. 5 b. Short crack

Figs. 5 a and b. Cracks in hardened and tempered axle journals of manganese-vanadium-steel, cross sections, etched in nital. 100 x

the steel containers by hydrogen decarburisation must be taken into consideration from temperatures of little above 200° C on upwards [6]) [7]).

The location of the decarburisation reaction during hydrogen attack is not the surface of the steel but any void, maybe a grain boundary, in the steel structure where the formated methane molecule can precipitate. The distances of carbon diffusion to the place of reaction are consequently very short ones. By the precipitation of methane the steel structure is burst apart. While without that the hydrogen penetrates rapidly into the

steel by diffusion, it is now still easier for it to enter and also made possible to remove the reaction products.

Corresponding to the altered reaction mechanism also the appearance of hydrogen decarburisation or hydrogen attack is completely different from that of decarburisation by oxydizing gases. The first thing to become visible is the bursting open of grain boundaries. The removal of a small amount of carbon is already sufficient. As an example Figs. 9 a to c show the structure of a soft iron with 0.01 % C before and after the action of hydrogen under 300 atm at 400 and 600° C [8]).

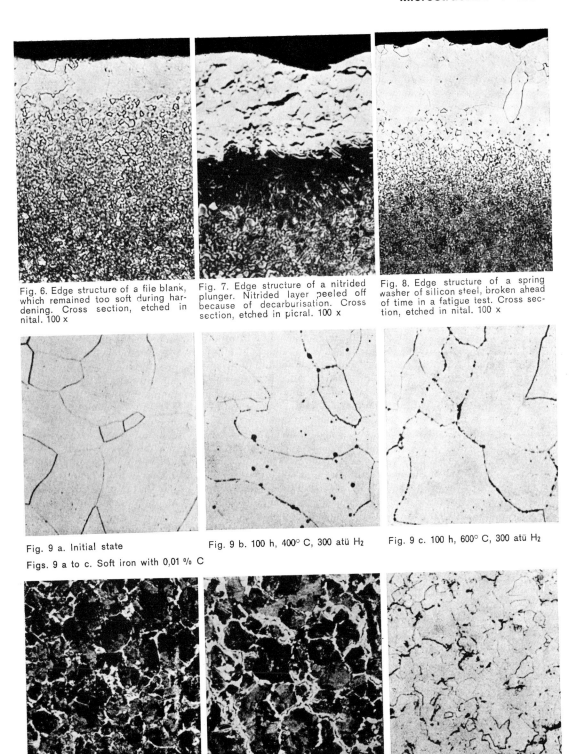

Fig. 6. Edge structure of a file blank, which remained too soft during hardening. Cross section, etched in nital. 100 x

Fig. 7. Edge structure of a nitrided plunger. Nitrided layer peeled off because of decarburisation. Cross section, etched in picral. 100 x

Fig. 8. Edge structure of a spring washer of silicon steel, broken ahead of time in a fatigue test. Cross section, etched in nital. 100 x

Fig. 9 a. Initial state

Fig. 9 b. 100 h, 400° C, 300 atü H_2

Fig. 9 c. 100 h, 600° C, 300 atü H_2

Figs. 9 a to c. Soft iron with 0,01 % C

Fig. 10 a. Initial state

Fig. 10 b. 100 h, 400° C, 300 atü H_2

Fig. 10 c. 100 h, 600° C, 300 atü H_2

Figs. 10 a to c. Steel with 0,45 % C

Figs. 9 and 10. Change in structure by hydrogen attack, etched in nital. 200 x

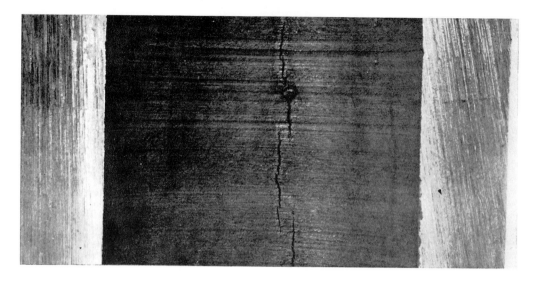

Fig. 11. High pressure tube burst by hydrogen attack. Inside view. 1 x

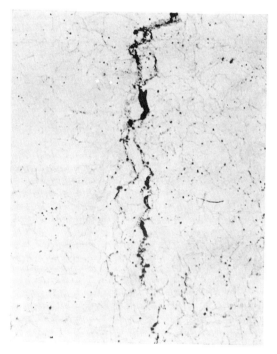

Fig. 12 a. Unetched

Fig. 12 b. Etched in picral

Figs. 12 a and b. Intergranular crack. 100 x

From the structure of steels with a larger content af carbon attacked by hydrogen it can be recognized how the decarburisation advances from the grain boundaries to the core of the grains. Figurès 10 a to c show that at a steel with 0.45 % C. The hydrogen attack favourably follows the dendrite axes with little segregation or primary bands. The initiations of such a hydrogen attack cannot always be detected easily by

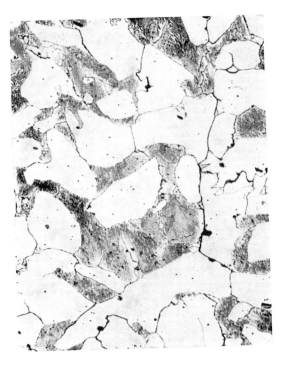

Fig. 13. Structure in the middle of the tube wall. 200 x

Fig. 14. Structure 3 mm away from the outside wall. 200 x

Figs. 12 to 14. Cross sections, etched in picral

Fig. 15 a. Initial state

Fig. 15 b. 100 h, 400° C, 300 atü H₂

Fig. 15 c. 100 h, 600° C, 300 atü H₂

Figs. 15 a to c. Change of the fracture of notched-bar impact specimen of a steel with 0,24 % C by hydrogen attack. 2 x

metallographic means, particularly when the temperature was low and thus the carbon diffusion slow.

Figure 11 shows the inside view of a high pressure tube of 110 x 70 mm diameter. For five years it had been exposed to hydrogen attack of max. 260 atm. at max. 280° C and had become leak by cracking up in longitudinal direction. After bursting open the crack faces had that mat grey appearance

which is significant for hydrogen attack. As proved from Figs. 12 a and b the crack is mostly intergranular. Already the unetched microsection (Fig. 12 a) reveals that the cohesion of the grains has loosened in the whole crack region. As could be expected from the low operation temperature the steel is only slightly decarburised. The hydrogen attack has greatly advanced. In the middle of the tube wall the structure is still heavily destroyed (Fig. 13) and even 3 mm away from the outer face the bursting open of grain boundaries can be recognized (Fig. 14). By loosening the structure the strength is decreased and all toughness data drop to a fraction of their original values. Fractures adopt the appearance of wood fibre structure and mat grey colour (Fig. 15). Similar to pickle attack, hydrogen attack can sometimes be externally perceived by the formation of blowholes which, however, are not filled with molecular hydrogen but with methane.

Hydrogen attack can be prohibited by alloying the steel with elements which formate steady and hardly soluble carbides like chromium, molybdenum, vanadium or titanium [5].

[1] W. OELSEN, K.-H. SAUER, G. NAUMANN, Zur Entkohlung von Stählen in zundernden und nicht zundernden Gasen. Landesamt für Forschung des Landes Nordrhein-Westfalen Jahrbuch 1969, 411/67. Westdeutscher Verlag, Köln und Opladen.

[2] F. K. NAUMANN, F. SPIES, Prakt. Metallographie 6 (1969) 447/456

[3] F. K. NAUMANN, F. SPIES, Prakt. Metallographie 6 (1969) 579/584

[4] F. K. NAUMANN, F. SPIES, Prakt. Metallographie 7 (1970) 155/159

[5] F. K. NAUMANN, Stahl und Eisen 58 (1938) 1239/1249

[6] I. CLASS, Stahl und Eisen 80 (1960) 1117/1135

[7] G. A. NELSON, Stahl und Eisen 80 (1960) 1134

[8] F. K. NAUMANN, Stahl und Eisen 57 (1937) 889/899

Cracked Eccentric Camshaft

Friedrich Karl Naumann and Ferdinand Spies
Max-Planck-Institut für Eisenforschung
Düsseldorf

During dismantling of an eccentric camshaft of 340 mm diameter that had worked for about $1/2$ year with a total of 450,000 load reversals, it was found that it had cracked on both sides of the eccentric cam. The shaft was made of chromium-molybdenum alloy steel 34 CrMo4 (material No. 1.7220) according to DIN 17200 and was heat treated to a strength of 69 kp/mm^2.

First a section of the one transition between the shaft and the cam lobe was sent in for examination. The piece showed two circumferential parallel cracks in the fillet which

Fig. 1. Fatigue fractures in fillet between shaft and side of cam lobe after application of magnaflux method. 1 ×

Fig. 2. Grinding checks on cam lobe side, after application of magnaflux method. 1 ×

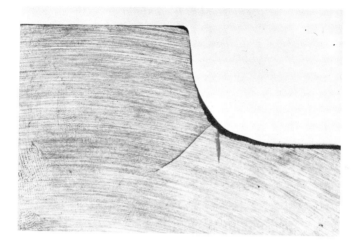

Fig. 3. Fatigue fractures in cross section, after application of magnaflux method. 1 ×

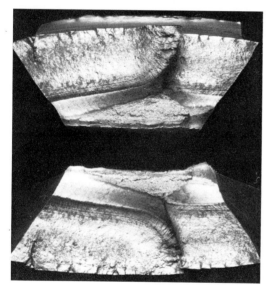

Fig. 4. Fatigue fractures in fillet breakout. 1 ×

apparently initiated independently of each other (Fig. 1). A number of cracks could also be observed running vertically to these in the adjacent zone of the cam lobe. These were recognized as grinding checks (Fig. 2).

Subsequently the remaining piece was viewed in the plant of the sender and it was found that a crack had occurred at this side running along half the circumference. At the beginning or at the branches it ran at an angle of about 30° against the circumference. This showed that it was a torsion fracture that was

deflected into the direction of circumference by the dimensional effect of the cross sectional transition. No bend fractures could be expected at this place because the incipient cracks were located at the compression side of the shaft. The finish of the fillet was good. Grinding checks could again be observed at the side of the cam lobe, i.e. in the characteristic crazing pattern.

The circumferential cracks were 15 to 16 mm deep as measured at cross sections (Fig. 3). As expected, they were fatigue cracks each

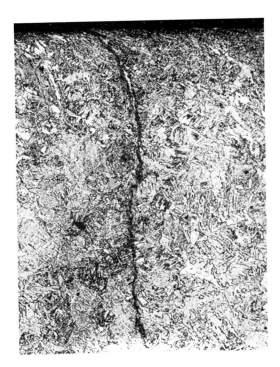

Fig. 5. Fatigue fracture with penetration of bearing bronze. Unetched section. 500 ×

Fig. 6. Grinding check on cam lobe side. Etch: Nital. 100 ×

Fig. 7. Cold deformation and grinding martensite at surface of cam lobe side. Etch: Nital. 200 ×

with several points of origin which was confirmed by breaking them open (Fig. 4). The crack propagations were discolored red by bearing metal. In microsections was to be seen that bearing bronze had penetrated not only into the incipient fatigue cracks but also into the finest branches of the grinding checks in the cam lobe (Fig. 5). The peripheral structure of the lobeside showed signs of strong cold deformation (Fig. 6). In some places a thin martensite layer had formed through heat caused by friction (Fig. 7). The shaft therefore ran hot. There were no material defects.

Therefore the shaft was probably overstressed by torsion forces. The presence of surface checks on both sides of the cam lobe that were filled with bearing metal proved that overstressing occurred through galling of the end faces of the bearing liners.

Destroyed Needle Bearing of a Packing Machine

Friedrich Karl Naumann and Ferdinand Spies

Max-Planck-Institut für Eisenforschung
Düsseldorf

A needle bearing from a filling- and sealing machine for milk cartons became unusable due to corrosion and fracture of a ring after only 4 weeks' operation of the machine in a Finnish milk packing plant. These bearings are subject to corrosion by water condensates in this type of environment because of constant temperature changes and they normally are replaced after 8 months. The bearings are lubricated by a molybdenum sulfide paste. The operation conditions for this machine differ from others with longer bearing life in that temperature differences are more pronounced and also thicker paper is compressed. Therefore it needed to be investigated whether corrosion in this case was accelerated by fracturing or break-away of the surface.

The needles and rings of the bearing were covered with a thick black-brownish crust, especially on the inside. After removal of the crust, corrosion pits were seen especially on the inside of the inner ring and on the needles. The outer ring showed crack initiation on the stressed side (Fig. 1, below).

Chemical analysis of the crust gave the following values (in wt.%):

C	S
7,37	1,65
Mo	Fe
2,20	69,6

Therefore it consisted principally of rust and contained Mo and S as constituents of the lubricant, as well as considerable amounts of carbon. Thus the bearing was lubricated not only with molybdenum sulfide paste but also evidently at times with graphite.

For metallographic examination the end face on both sides of the rings was ground off till the groove was removed. A longitudinal section was also made. The etched sections of the outer ring are shown in Fig. 2. The rings were case hardened whereby the acutely angled edges were overcarburized at the fold of the outer ring and were also overhardened (Fig. 3). But the cracks propagated from the inside and therefore had no connection with the overcarburized edge. Their presence pointed to overstressing by compression of the bearing. The sections of the inner ring and the needles showed deep corrosion pits (Figs. 4 and 5).

Judging by their structure the needles probably consisted of ball bearing steel. They showed corroded initial cracks of the pitting type, i.e. shear-fatigue fractures due to excessive surface pressure (Figs. 6 and 7). The

Fig. 1. Outer surface of fractured outer ring. 1 × .

needles too were overstressed by compression. Whether or not the initial pressure was too high or whether the failure was caused later by rolling of the corrosion products into the material, could not be definitely decided. It seems that the higher pressure necessary for the pressing of the thicker paper accele- rated the corrosion which lead to the crack initiations of the parts and possibly also to impaired lubrication.

The machine manufacturer has therefore switched to bearings with shells of a complex bronze.

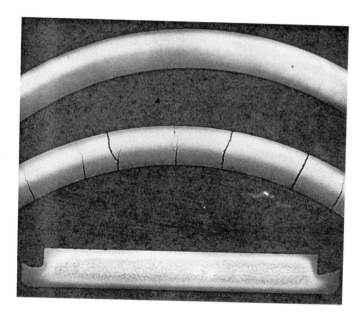

Fig. 2. Sections through outer bearing ring. Etch: Picral. 3 ×.

free edge, cross section

clamped edge, cross section

longitudinal section

Fig. 3. Structure of outer bearing ring in overcarburized edge of groove. Etch: Picral. 100 ×.

Fig. 4. Corrosion pitting in bearing ring. Unetched longitudinal section. 100 ×.

Fig. 5. Corrosion pitting in needle. Unetched transverse section. 100 ×.

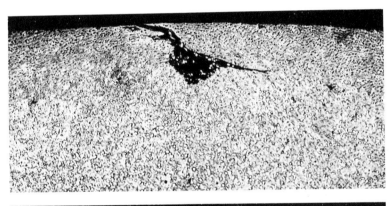

Fig. 6

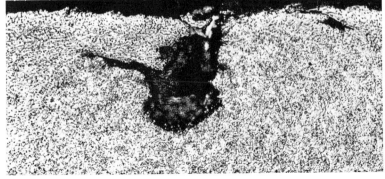

Fig. 7

Figs. 6 and 7. Partially corroded pitting in needles. Transverse section, etch: Nital. 500 ×.

Fractures of Flat Wire Conveyor Belt Links of Glass Annealing Furnaces

Friedrich Karl Naumann and Ferdinand Spies

Max-Planck-Institut für Eisenforschung

Düsseldorf

The cross bars of conveyor belt links that served to transport glass containers through a stress relief furnace fractured in many cases. They consisted of wires of 5 mm diameter made of low carbon Siemens-Martin steel, while the interwoven longitudinal bars were made of strip steel of 4 x 2 mm². The furnace temperature was said to be 500°C. After leaving the furnace the endless belt was conducted over a roller of 600 mm diameter and returned to the furnace via a tension roller and several redirect rollers of 200 to 300 mm diameter.

Figure 1 shows a number of cross bars. In addition to the fractures they also showed many more or less advanced cracks. These occurred in the circumferential grooves that recurred at regular intervals. The fractures were abraded and oxidized. They also could possibly be fatigue fractures.

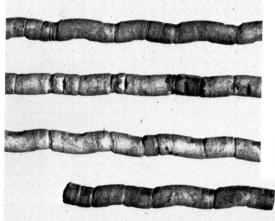

Fig. 1. View of cross bars of conveyor belt. 1 x

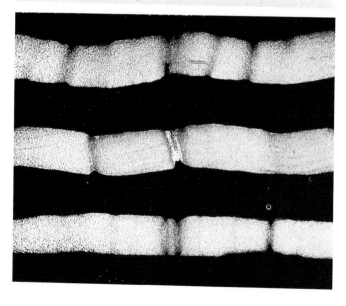

Fig. 2. Longitudinal sections through cross bars. Etch: Nital. 3 x

From the etched longitudinal sections (Fig. 2) it could be seen that the wire surface and the crack walls were oxidized (Fig. 3). Under the ring-shaped notches and at the fracture points narrow zones could be observed that were in part lighter and in part darker and had a changed microstructure. As can be seen from Fig. 4 these zones were cold deformed and had partly recrystallized into coarse grains in the center under the notches. These narrow, thoroughly deformed areas could only have been caused by shear or torsion with short gripping length. Torsion marks could also be observed externally at the surfaces. It may also be possible that the cross bars were turned back and forth during

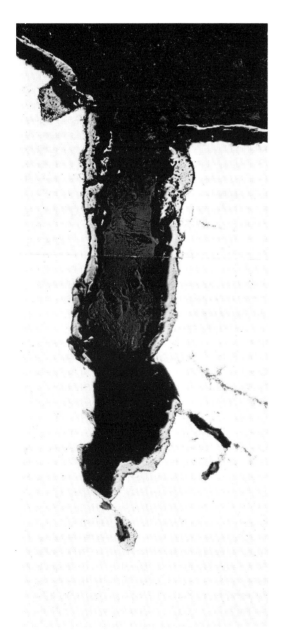

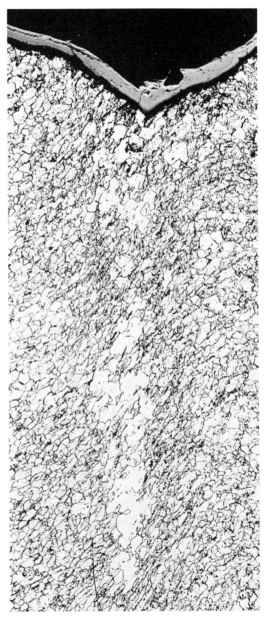

Fig. 3. Oxidized crack in cross bar. Longitudinal section, unetched. 100 x

Fig. 4. Cold deformation and recrystallization under a groove. Longitudinal section, etch: Nital. 100 x

movement of the belt around the redirect rollers whereby the strained strip steel windings acted as damping heads. The fracture probably was induced by the pressing-in or abrading of the sharp steel band edges into the surface of the cross bars. Torsion fatigue fractures may be started from these notches. Relaxation then contributed positively through recovery and recrystallization.

These assumptions of the fracture cause are supported by the observation of the sender, that such damage occurs less frequently in round wire conveyor belt links because the round wire neither impresses so sharply nor abrades against the cross bars and it also exerts less torsion than the flat wire. With this oberservation by the user the remedy is at hand.

Working Roll with Shell-Shaped Fractures

Friedrich Karl Naumann and Ferdinand Spies

Max-Planck-Institut für Eisenforschung
Düsseldorf

A working roll of 210 mm diameter and 500 mm face length was examined because of shell-shaped fractures. The roll consisted of chromium steel of the following composition in wt.-%:

C	Si	Mn	P	S	Cr
0,83	0,23	0,20	0,020	0,007	1,60

The chromium content was low for a roll of this diameter.

The fracture of the roll is shown in Fig. 1. The crack origin is designated by an arrow. It was located approx. 10 mm under the roll face. From this point the fracture in form of a narrow strip, whose special structure can be seen better at a higher magnification in Fig. 2, runs concentrically around almost two-thirds of the circumference and at various points exits at the surface obliquely from the strip. The fracture had a fine-grained structure. Figure 3 shows the roll after etching of the surface with 5% Nital. The dark strips are a tempered structure and prove that the roll became hot in places. This probably happened through sliding of the material to be rolled[1]. But it is also entirely possible that the hot spots occurred at least in part after fracturing through polishing of break-away splinters. As can be seen from the transverse section (Fig. 4) and the hardness-depth curve (Fig. 5) the roll was hardened to a depth of 18 to 20 mm. The peripheral structure consisted

Fig. 1. 210 mm diameter roll with fragments. Fracture origin designated by arrow.

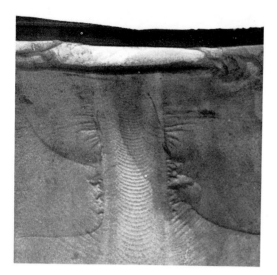

Fig. 2. Structure of strip located approx. 10 mm under face from which fracture propagated. Approx. 1/2 ×

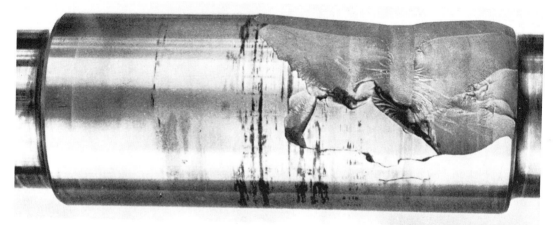

Fig. 3. Roll after etching of face with 5% Nital.

Fig. 4. Transverse section through plane of fracture origin. Etch: Copper ammonium chloride solution according to Heyn.

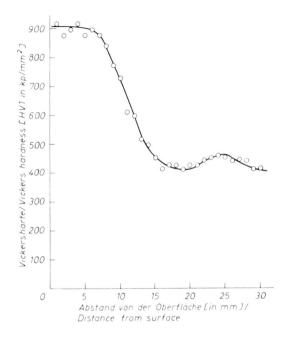

Fig. 5. Hardness course under roll face

of very fine acicular martensite with fine spherical carbide inclusions (Fig. 6) up to a depth of 5—6 mm. The martensite indicated no signs of tempering treatment. Adjoining the purely martensitic peripheral zone was a transition zone with increasing amount of bainite and pearlite from the exterior to the interior, (Fig. 7) that led into the pearlitic core (Fig. 8).

Surface hardness (HV1) of 900 kp/mm^2 was exceptionally high corresponding to the martensitic peripheral structure. An untempered piece with such a thick cross section and a hardened peripheral zone with such high hardness must have high residual stresses that culminate in the transition zone. Therefore it must be very sensitive against additional stresses, be these of a mechanical or thermal nature. This contributed to the fragmenting of the roll face.

[1] Vgl. F. K. NAUMANN, F. SPIES, Prakt. Metallographie 5 (1968) 647/651

(Continued on the next page)

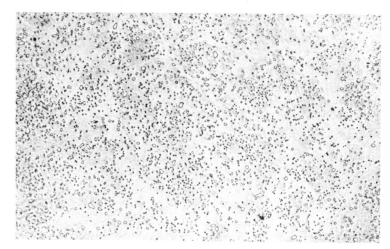

Fig. 6. Martensitic peripheral zone (up to 6 mm depth)

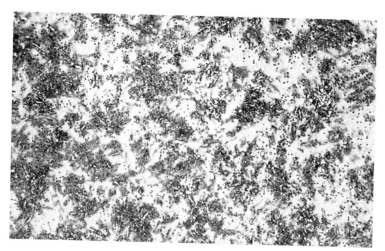

Fig. 7. Transition zone (martensite with bainite)

Fig. 8. Pearlitic core structure

Figs. 6 to 8. Microstructure of roll. Transverse section, etch: Picral, 500 ×

Galvanised Cable Damaged by Localised Heating

Elfriede Waldl

Höhere Technische Bundeslehranstalt
Salzburg

Two sections of a galvanised cable 10.5 A 160 GR ÖNORM M 9533 (round stranded cable of normal type, h + 6, Langslay, right-handed) were brought for examination. One had a ca. 100 mm long blackish-brown tarnished zone obviously caused by localised heating at one end, inside which the hemp core was missing, and the other corresponded to the original condition of the cable. The cause of the damage is unknown. Figure 1 shows one end of the faulty section with the site of the damage. About a third of the wires had fractured and the rest had been cut. All were tensile fractures with a relatively high degree of necking (Fig. 4).

Seven wires were taken from three different strands of the damaged section and six wires from three different strands of the undamaged section and their tensile strengths measured. Whereas the average tensile strength of the wires from the undamaged section was 161.5 ± 1 kgf/mm², thus fulfilling the specification, the tensile strength of the wires from the damaged section varied between wide limits (between 81.5 and 139.7 kgf/mm²) in no case reaching the required value of ca. 160 kgf/mm². The damaged section thus had a tensile strength from 26 % − 50 % lower than that of the undamaged section. Figure 2 shows the stress-strain curve for a sound wire and Fig. 3 that for a damaged wire.

Longitudinal sections from ruptured and not

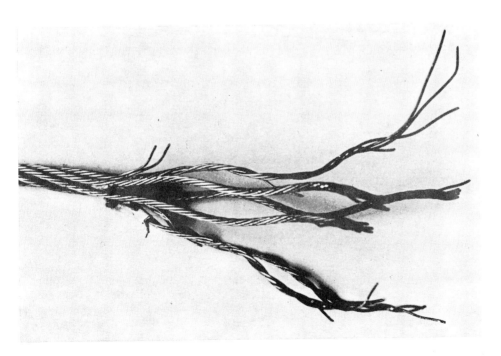

Fig. 1. Damaged area. 3 x

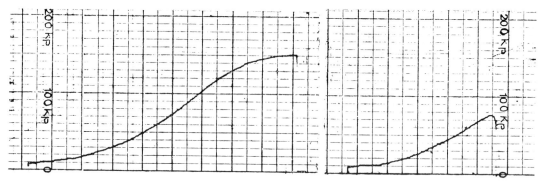

Fig. 2. Stress-strain curve of the undamaged wire, σ_B = 161 kgf/mm²

Fig. 3. Stress-strain curve of damaged wire, σ_B = 81.5 kgf/mm²

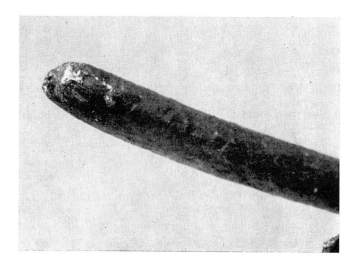

Fig. 4. Fractured surface of a ruptured wire. 5 x

ruptured wires from the damaged section and wires from the undamaged section were prepared for metallographic examination. Figure 5 shows the microstructure of the wires in the original condition as found in the undamaged wires and in the untarnished portion of the damaged wires. It consists of very long elongated grains of fine lamellar pearlite with little ferrite: the typical patent structure of a cold drawn wire. In the tarnished portion, however, the structure has changed fundamentally (Fig. 6). It consists of granular pearlite without grain elongation, which indicates that it has been soft annealed, thus explaining the reduction in strength. The edge zone has been heavily decarburized (Figs. 7 to 10).

The zinc layer has apparently reacted with the base material to form a solid solution or a compound. This takes on a dark colour when etched with alcoholic picric acid (Fig. 8) and is significantly harder than the core structure. The microhardness of the decarburized layer was about 140 kgf/mm², that of the core about 197 kgf/mm² and that of the solid solution varied between 266 kgf/mm² and 434 kgf/mm² at a load of 10 gf.

The outer zone could be a solid solution

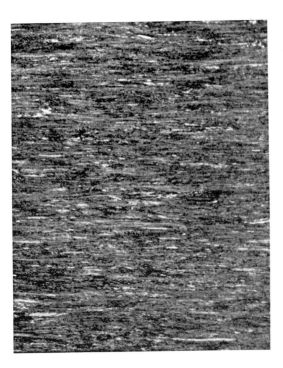

Fig. 5. Microstructure of the core of an undamaged wire. Etched in alc. picric acid. 500 x

Fig. 6. Microstructure of the core in the annealed region of a damaged wire. Etched in alc. picric acid. 500 x

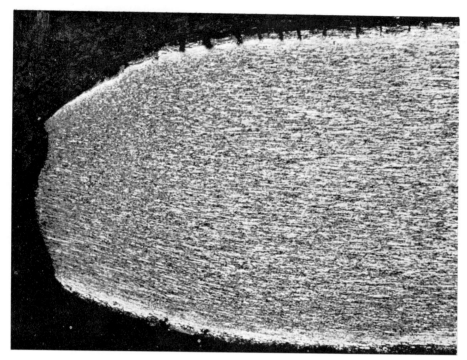

Fig. 7. 100 x

Fig. 8. 500 x

Figs. 7 and 8. Longitudinal section through the fractured surface of a wire. Etched in alc. picric acid

containing carbon or the carbide Fe_3ZnC. This layer is obviously brittle and has cracked in many places under the load carried by the cable. The notches so formed have initiated the fracture of the wires and reduced the strain to fracture, as can be seen from Fig. 3.

The cause of the localised heating is unknown. It can only be concluded from the investigation that the temperature did not exceed the Ac_3 point of the wire material, which should be about 750° C, and that the heating lasted a fairly long time.

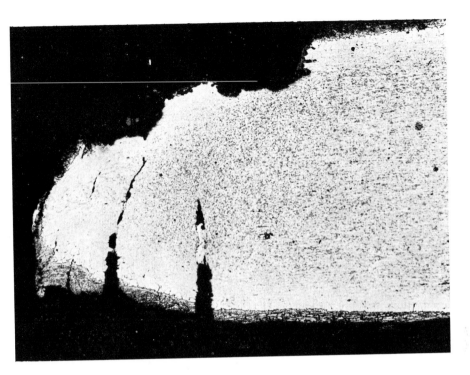

Fig. 9. 100 x

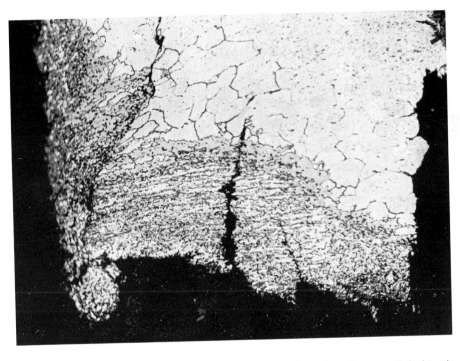

Fig. 10. 500 x

Figs. 9 and 10. Longitudinal section through the fractured surface of another wire. Etched in alc. picric acid

Ruptured Stainless Steel Heater Tube

Fulmer Research Institute Ltd.
Buckinghamshire, England

Three samples from a ruptured 316 stainless steel tube were examined. The samples are shown in Fig. 1. The unmarked sample was taken from the centre of the rupture where the tube thinned, bulged, spread and became partially flattened. Sample 1 was taken from the edge of the fire adjacent to the unmarked sample. Sample 2 was situated about 400 mm from the centre of the fire.

The tube, 114 mm o. d., wall thickness 8.00 mm, with 13 mm thick 321 stainless steel finns welded to the outer surface of the tube, was part of a heater through which sour gas, containing methane plus H_2S and CO, passed at 1150 p. s. i. g. The sour gas was heated to 600° F by burners playing on the outside of the tube burning "sweet" gas plus air.

Approximately 2 hr prior to rupture of the

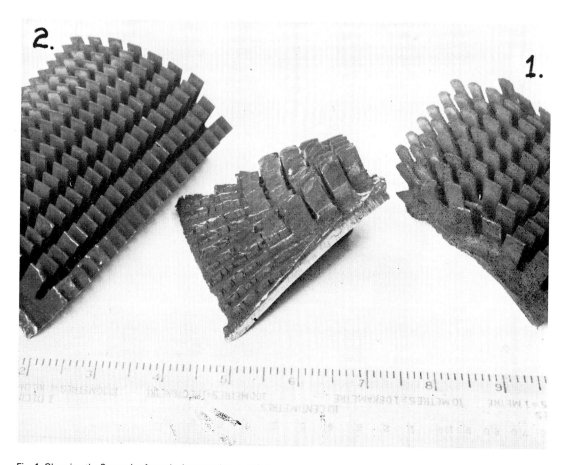

Fig. 1. Showing the 3 samples from the heater tube. ~ 1/2 ×

Fig. 2. Cavitation and thinning of a section from the unmarked sample. ~5×

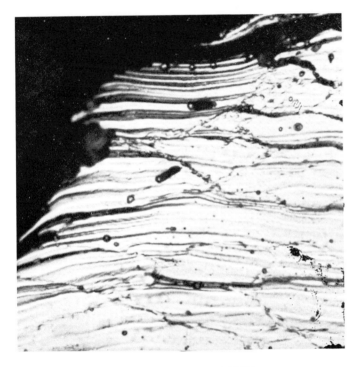

Fig. 3. Area of section in Fig. 2, at higher magnification and etched in 10% oxalic acid, showing ductile deformation of austenite grains near the fracture surface. 500×

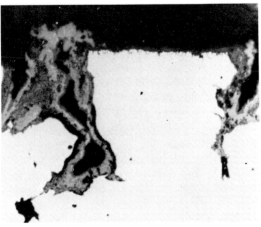

Fig. 4. Outside surface of unmarked sample showing fingerlike corrosion attack. Unetched. 250×

tube and the consequent fire, a temperature recorder showed that the temperature of the sour gas exceeded 800° F, which is the top position of the recorder. The maximum temperature reached was not therefore known, but the period lasted approximately 30 min before the sour gas exit temperature returned to 600° F. At the time of rupture the tube was operating at its normal temperature and pressure.

The wall thickness of the unmarked sample had been reduced to a chisel edge along the fracture line, which ran in a direction parallel to the major axis of the tube. The wall thickness of sample 1 had also been reduced but not so with sample 2. The outer surfaces of all samples had a rusty oxide appearance. The inner surfaces of samples 1 and 2 were covered with a black, slightly glazed, adhering scale whereas the inner surface of the un-

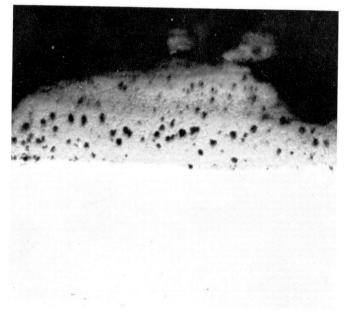

Fig. 5. Inside surface of sample 2 showing general attack. Unetched. 500 ×

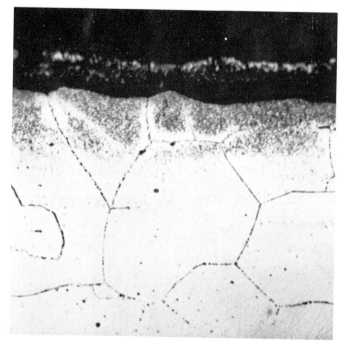

Fig. 6. Inside surface of sample 2 showing general attack and precipitation effects. Etched with 10% oxalic acid. 500 ×

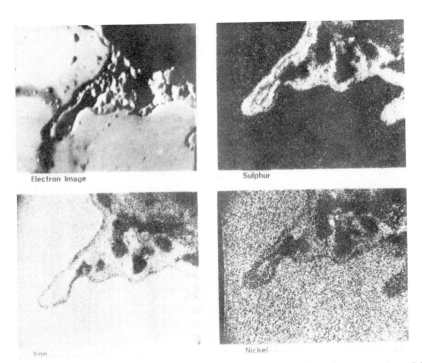

Electron Image

Sulphur

Iron

Nickel

Fig. 7. Showing electron and X-ray images of finger-like corrosion from the outer surface of the unmarked sample. 250 ×

marked sample was only discoloured black without any scale.

Metallographic examination showed that the basic microstructures of all samples were similar, consisting of equiaxed austenite with occasional areas of a second phase believed to be ferrite. The grain size of all sections was also similar varying between the ASTM grain size numbers 4—5.

Figure 2 shows a section through the unmarked sample and exhibits cavitation at and near to the fracture surface together with local thinning. Figure 3 illustrates the fracture region at higher magnification and shows ductile deformation of the austenite grains.

The inner and outer surfaces of all samples showed evidence of corrosive attack. The most aggressive attack occurred on the outside surface of the unmarked sample and took the form of local finger-like intrusions, containing complex elevated temperature corrosion products in eutectic form as shown in Fig. 4. The maximum depth of attack at the inside surface was not greater than 0.2 mm whereas the outside surface had been attacked to a depth of 0.3 mm. Some grain boundary carbide precipitation along the inner surface had also occurred.

The corrosion at the inner and outer surfaces of samples 1 and 2 was mainly of a general "layer" type (see Fig. 5) similar in appearance to the corrosion product in the corrosion fingers. There was also some carbide precipitation within the grains and at the grain boundaries as shown in Fig. 6.

The results of electron probe microanalysis of a section taken from the unmarked sample are illustrated in the form of electron and X-ray images in Fig. 7. These results showed that the corrosion products contain sulphur with iron together with nickel to a lesser extent. There was, however, one area high in nickel and sulphur present at the centre top region of the nickel X-ray image.

Local thinning, cavitation and ductile deformation markings associated with the unmatched sample taken from the centre of the fire showed that the tube ruptured as a result of overheating. The corrosion product and precipitation effects at the inner and outer surfaces were considered to have formed during the overheating and subsequent fire periods and not under normal service conditions.

The failure mechanism envisaged was that overheating occurred (through reasons impossible to establish from the investigation) resulting in loss of strength of the tube material which could not then sustain the normal running conditions. Sulphidation could have occurred during the overheating period thereby weakening the tube just prior to rupture.

Analysis has shown the corrosion products to be complex containing iron, nickel and sulphur. The form of the corrosion product, particularly in the corrosion fingers (see Fig. 4) has a eutectic appearance possibly either nickel sulphide or iron sulphide. The nickel sulphide eutectic system forms at a lower temperature (about 1200° F) than the iron sulphide system ($\sim$1800° F).

Summary

Overheating whilst the temperature recorder was off the chart caused severe loss of tube strength, resulting in ductile rupture. The minimum overheating temperature could be deduced at around 1200° F due to the presence of a eutectic observed metallographically within the surface corrosion products.

Damages in Gears in the Transmission System of Heavy Duty Tracked Vehicles

P. K. Chatterjee

Controllerate of Inspection
Ichapur-Nawabganj, 24 Parganas, West Bengal

At one time there were instances of failure in the teeth of a spur gear in the transmission system of heavy duty tracked vehicles. The gears are made from a case hardening nickel-chromium-molybdenum type steel with the following nominal composition in wt.-%:

Carbon	0.20 max.
Manganese	0.50/1.00
Nickel	1.00/1.50
Chromium	0.75/1.25
Molybdenum	0.08/0.15

Case carburized to 1 mm depth

Hardness of case 58 to 62 Rockwell C (708 to 802 VPN)

The defects were in the nature of seizure on the involute profile. Transverse sections cut through these defective zones showed a whitish colour due to slow etching constituents in contrast to the dark-etching case hardened profile of the gear teeth. A typical defective area is shown in Fig. 1. In Figs. 2 a and b can be seen the marks due to the rolling effect of the gears at the lines of contact of the mating gears in the affected region. The microhardness in the damaged portions of the case (white in colour) and the unaffected portion of the case (dark etching) were 940 and 724 HV respectively. The white area is obviously the result of a change in the microstructure due to reformation of austenite and martensite by the

Fig. 1. Enlarged view of a typical damaged tooth of the spur gear. The damaged area of the tooth (white in colour) is shown by an arrow. 7 x

Fig. 2 a. Defects in the damaged area of the tooth in Fig. 1. 28 x

Fig. 2 b. Enlarged view of the defects shown in Fig. 2 a, rolling effects on gear contour. 100 x

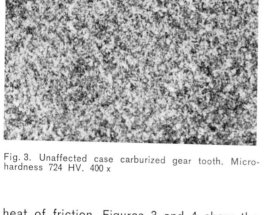

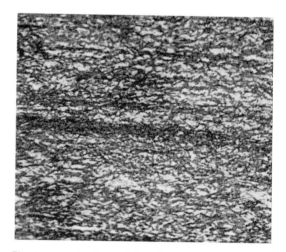

Fig. 3. Unaffected case carburized gear tooth. Microhardness 724 HV. 400 x

Fig. 4. Microstructure of the damaged area, reformed austenite and martensite. Microhardness 924 HV. 400 x

heat of friction. Figures 3 and 4 show the original structure of the case and the structure as modified by the heat of friction, respectively.

Scrutiny of the transmission system showed there might arise possibilities of choking in the line for the lubricating oil. Such eventualities would, of course, cause seizure of the gears and damage. The incidence of such defects stopped after corrective measures were taken.

In this particular instance, corrective measures could be taken by visual observations of the defects. The microscopic examination is of value in assessing the structural changes likely to arise in conditions of extreme pressure and failure of lubrication. Incidentally it also delineated the flow of material in surface layers of gears under conditions of high pressure and failure of lubrication.

Failure of a Flange from a High Pressure Feeder Plant

Ladislav Kosec, Franc Vodopivec and Bogomir Wolf

Metallurgisches Institut
Ljubljana, Jugoslavia

The flanged bearing bush carrying the drive shaft of a feed pump suddenly fractured after about two years' service. The play between the shaft and the internal wall was ca. 0.1 mm. The pure water flowing through this gap can be considered as a lubricant. The lubrication broke down as a result of an unknown fault and the drive shaft came into direct contact with the internal wall of the flange.

On fracture, the flange fell into several pieces (Fig. 1). Large cracks were observed on the individual sections originating on the inner side of the flange (Fig. 2). The fracture surfaces were coated with a thin layer of magnetite scale. Wide and narrow strips of varying thickness of a smeared extraneous metal were visible on the inner surface of the flange (Fig. 3). Similar information was provided by macrographs of transverse (Fig. 5) and longitudinal sections (Fig. 4) of the flange. Part of the smeared metal on the flange is visible in Fig. 4. It can be seen to contain grains of scale. Macroscopic etching reveals numerous fine cracks not visible to the naked eye (Fig. 5). Figure 5 also shows

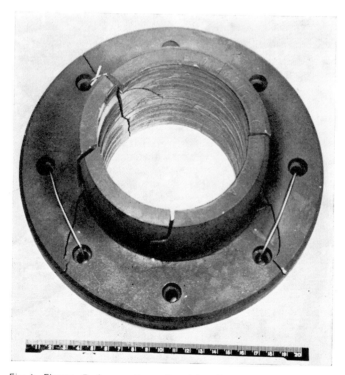

Fig. 1. Flange. Broken sections pieced together

Fig. 2. Individual pieces of the ruptured flange

Fig. 3. Inner side of a piece of broken flange. Bands of another smeared metal

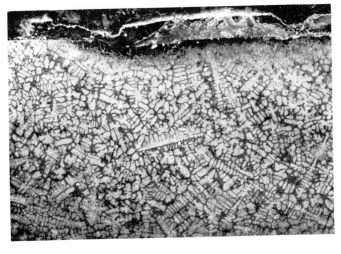

Fig. 4. Cross section through flange. Smeared metal on the inner wall. Etched ($CuSO_4$ + HCl + H_2O). 15 x

Fig. 5. Cross section on the inner side of the flange. Cracks and broad heat affected zone. Etched (FeCl₃ + HCl + ethanol). 2.6 x

a zone approximately 1.5 cm broad which has undergone structural changes as a result of the raised temperature. The chemical composition (wt. %) C 1.54; Si 0.88; Mn 0.51; P 0.025; S 0.026; Cr 18.2; Mo 0.11 and Ni 0.36 is normal for high chromium ledeburitic cast steel, which is corrosion and wear resistant as well as refractory.

The microstructure shows that the cast material is relatively pure and that all the inclusions are oxidic (essentially chromium oxide). The original microstructure outside the heat affected zone consists of ferrite with finely dispersed carbides. The ferrite grains are surrounded by ledeburitic eutec-

tic. Such a structure is normal for this alloy. In the part of the flange still possessing the original cast structure the cracks run mainly across the ledeburite eutectic i.e. across the brittle structural component of the cast material (Fig. 6). The microhardness of the ferrite grains with finely precipitated carbides is 350 to 370 kgf/mm² (HV) but of the ledeburitic eutectic ca 530 kgf/mm² (HV).

The microstructure of an approx. 15 mm wide zone has changed as a result of heating to a high temperature and subsequent rapid cooling. The microstructure thus produced (Fig. 7) has a hardness of 620 to 690 kg/mm² (HV).

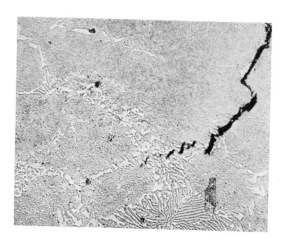

Fig. 6. Crack path in part of flange with unaltered microstructure. Etched (CuSO₄ + HCl + H₂O). 200 x

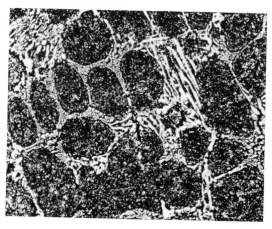

Fig. 7. Martensite with unaltered ledeburite network. Etched (FeCl₃ + HCl + ethanol). 100 x

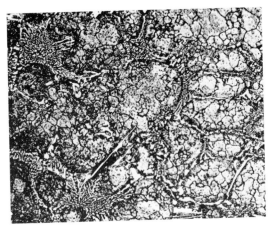

Fig. 8. Microstructure of heat affected zone. Left martensite (black) with ledeburite eutectic, right recrystallised structure. Etched (FeCl₃ + HCl + ethanol). 100 x

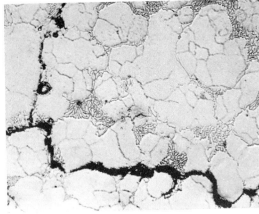

Fig. 10. Crack path on the inner surface of the flange in the heat affected zone. Etched (FeCl₃ + HCl + ethanol). 100 x

Fig. 9. Microstructure on the inner side of the flange in the heat affected zone; left austenite with ledeburite, right austenite with localised melting. Etched (FeCl₃ + HCl + ethanol). 100 x

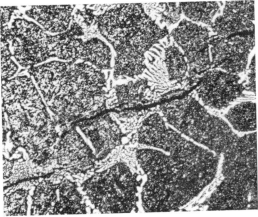

Fig. 11. Crack path in the heat affected zone. The crack runs partly through the ledeburite eutectic and partly through the martensite grains. Etched (FeCl₃ + HCl + ethanol). 100 x

In the direction of the surface where the temperatures were higher one observes the appearance of new grains (Figs. 8 and 9), solution of carbides and localised melting (Figs. 9 and 10). According to this the temperature may well have exceeded 1200 °C. This was confirmed by annealing experiments. The cracks no longer always follow the ledeburitic regions in the heat affected zone (Figs. 10 and 11).

The smeared metal is permeated with scale (Fig. 12) and in many places welded with the flange material (Fig. 13). In other places the join is broken by scale. A high temperature is necessary to achieve such a weld and a high pressure also unless one of the two partners is soft or molten.

The microstructure of the smeared metal consists partly of martensite and residual

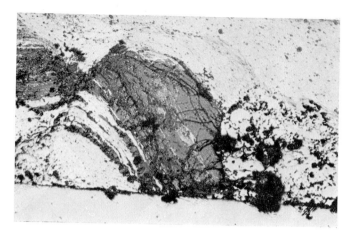

Fig. 12. Area on inner surface of flange smeared with another metal. Grey patches are scale. Unetched. 500 x

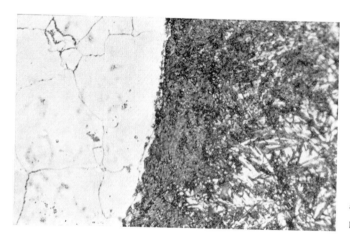

Fig. 13. Microstructure of contact zone between flange and smeared metal. Left austenite (flange), right martensite and residual austenite (smeared metal). Etched ($FeCl_3$ + HCl + ethanol). 500 x

austenite (Figs. 13 and 14). According to an investigation with the electron beam microprobe material with a high chromium content has penetrated relatively deeply into the smeared shaft material (Fig. 15a). Since the same areas are also enriched with oxygen (Fig. 15b) this may be chromium oxide from flange scale which has been kneaded into the welded shaft material.

On the basis of the phenomena described above the cause of the fracture of the flange can be sought in the following.

For unknown reasons the rotating shaft came into direct contact with the flange. Mechanical friction caused a rise in temperature on both contact surfaces. This mutual contact lasted long enough for the temperature in the contact zone to exceed 1200 °C at which the flange material became softened or molten. As a result of this, considerable structural changes took place on the inner wall of the flange. Thermal stresses and excessive mechanical loads due to smearing of the flange material then led to fracture of the flange.

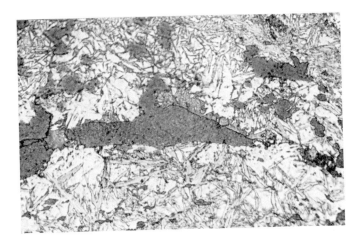

Fig. 14. Microstructure of metal smeared on flange: martensite and residual austenite. Large inclusions of scale are grey. Etched ($FeCl_3$ + HCl + ethanol). 500 x

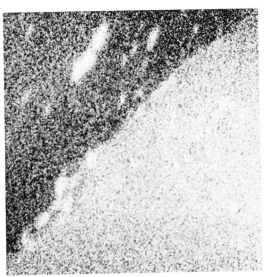

Fig. 15 a. Cr, Kχ

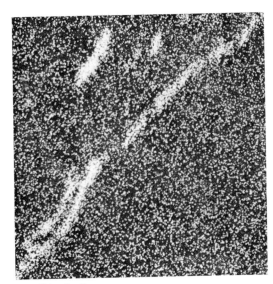

Fig. 15 b. O_2, Kχ

Figs. 15 a and b. Scanning micrographs of the area of contact between flange material and smeared metal. Top left smeared metal. 840 x

Section III:
Materials-Related Failures

A Broken Cross-Recessed Die Made from High Speed Tool Steel

Egon Kauczor

Werkstoffprüfamt der Freien und Hansestadt
Hamburg

Figure 1 shows a general view of the piece of the broken cross-recessed die under examination. For the purposes of the metallographic investigation, a longitudinal section was taken through the die on the opposite side to the fracture to include one arm of the cross.

Examination of the unetched, polished section for impurities revealed several coarse streaks of slag, one of which is illustrated in Fig. 2. The purity does not therefore correspond to the requirements set for a high speed tool steel of the given theoretical quality DMo 5.

After etching with 5 % nital the polished surface exhibited a pronounced, macroscopically easily visible, fibrous structure (Fig. 3). Microscopic examination revealed that this etch pattern was produced by marked segregation bands.

Figure 4 illustrates the structure of a longitudinal microsection through the fracture surface in the vicinity of the cross. The structure on the left side of the picture is finely acicular and was strongly attacked by the etching reagent. Adjacent to this and extending to the right side of the picture is a broad segregation band. Here the zone-

Fig. 1. General view of the fractured piece of a cross recessed die under examination. 2 x

Fig. 2. Slag streak in unetched longitudinal polished section. 100 x

wise less powerful attack of the etchant is the result of the accumulation of alloying elements as well as of carbides in the segregation bands. The increased concentration of alloying elements makes these bands vulnerable to overheating and excessive holding even during a heat treatment normal for this quality of steel, and causes abnormal hardening behaviour which has led to very coarse acicular structure and high hardness values in these zones. In particularly strongly developed segregation bands the steel can actually crack on hardening (coarse grain hardness cracks). The graphic representation of the results of low load hardness tests in Fig. 5 shows the differences in hardness between the segregation zone and the neighbouring fine acicular structure in Fig. 4.

Fig. 3. Structure near the fracture edge (etched with 2 % nital). 200 x

The very unfavourable structure for a high speed steel tool of these dimensions and subject to such stresses together with the low purity have at least favoured the fracture of the tool.

Fig. 4. Low load Vickers hardness test with test load of 500 gf in the segregation region shown in Fig. 3 (weaker etch than in Fig. 3). 200 x

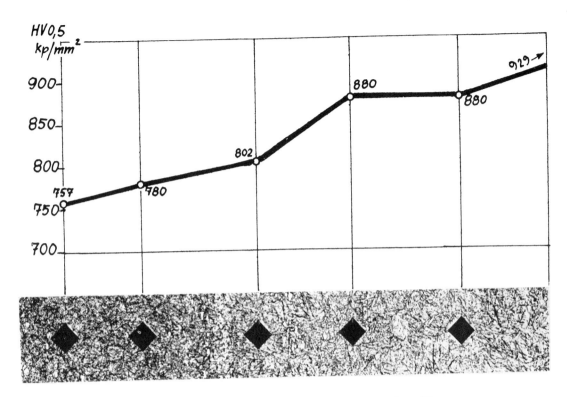

Fig. 5. Low load hardness values of the segregation band and neighbouring structure. 200 x

294

Steel Casting with Insufficient Strength Properties

Friedrich Karl Naumann and Ferdinand Spies
Max-Planck-Institut für Eisenforschung
Düsseldorf

In a steel foundry tensile and bend specimens of castings made in a 2 ton basic arc furnace showed at irregular intervals regions with coarse-grained fractures where the specimens broke prematurely, so that the specified strength and toughness values could not be reached.

After the slag was reduced by 90 % ferrosilicon powder and coke the 2-ton melt was deoxidized in a ladle with 3 kg aluminum, and was cast in wet sand molds. The pieces were stripped from the mold while red hot and were cooled in air and normalized at 900° C for 3 hours. The sender reported that tendency toward abnormal fractures rose with increasing melting temperature, but that it was absent in melts made in acid furnaces.

Several cast tensile specimens and some forcibly broken pieces of the flanges of armature yoks made of cast steel GS C 25 according to DIN 17 245 were investigated. Figures 1 a and b show two of the tensile specimen fractures. The cross section is not noticeably constricted at the point of fracture. The fractures have regions of coarse conchoidal structure. Such regions diminish the deformability the more, the larger they are in proportion to the total fracture cross section and the less favorably they are located with respect to the direction of principal stress. The disk of the flange broke almost completely in a coarse-conchoidal manner (Fig. 2). Based on these findings the advice was given to limit the aluminum addition to the steel as much as possible.

Fig. 1 a   Fig. 1 b

Figs. 1 a and b. Tensile test fractures. 2 x

Fig. 2. Fracture of a flange disk. 0.7 x

An analysis of the chips of the flange confirmed that the steel indeed contained considerable amounts of aluminum in addition to much nitrogen that is characteristic of basic arc furnace operation.

C %/o	Si %/o	Mn %/o	P %/o
0,22	0,28	0,89	0,021

	S %/o	Al %/o	N %/o
	0,013	0,115	0,014

A section was prepared parallel to the fracture plane of the flange disk for macroscopic metallographic examination. It was pickled with diluted hydrochloric acid. This caused the erosion of a coarse-meshed network that coincided with the primary grain boundaries in the globular solidified inner zone, but cut across the dendrites in the crystallized outer zone (Fig. 3).

Microsections were made from both zones. From them it could be seen that the networks eroded by pickling were coated with very fine precipitates that were located in the inner zone on the primary grain boundaries which

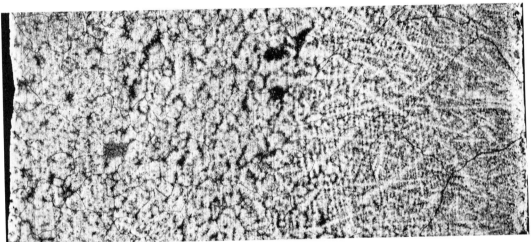

Fig. 3. Flange disk according to Fig. 2, section parallel to fracture plane. Etch: Diluted hydrochloric acid 1:1, 70° C. 2.5 x

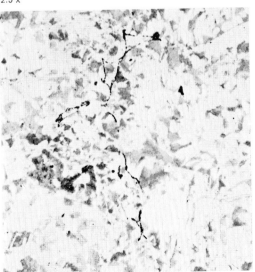

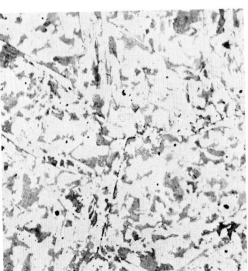

Fig. 4. Globular inner zone

Fig. 5. Transcrystallized peripheral zone

Figs. 4 and 5. Microstructure of disk according to Fig. 3. Etch: Picral. 100 x

Fig. 6. Precipitates on lattice planes and twin planes of former austenite. Etch: Picral. approx. 420 x

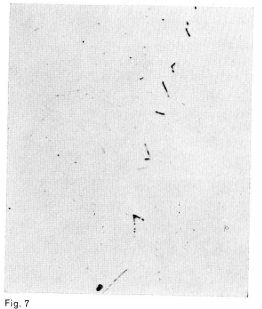

Fig. 7

Fig. 8

Figs. 7 and 8. Precipitates in grain boundaries and planes, unetched section of outer zone of flange disk. 1500 x

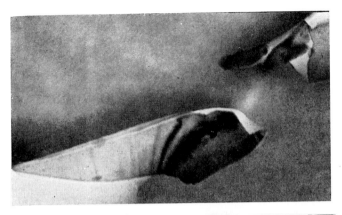

Fig. 9. Aluminium nitride precipitates in transmission lacquer replica. 40 000 x

Fig. 10. AIN platelets in carbon films stripped off fracture plane. 20 000 x

| 1 h | 8 h | 24 h | 1 h | 8 h | 24 h |

1100° C/Luft (air)   1200° C/Luft (air)

Fig. 11. Annealing tests with specimens from flange disk. 1 x (see additional designation at fractures)

were characterized by inclusions and micro-pores (Fig. 4). But in the crystallized outer zone they appeared in numerous cases as straight line fibers on the boundaries, or in certain lattice planes of the former austenitic grains (Fig. 5). This is shown even more distinctly in the higher magnification of Fig. 6. The precipitates have a rod-like or plate-like shape and they either have grown into the lattice plane if they originated at the grain boundaries, or they are located along the twin planes of the austenite (Fig. 8).

The optical examination was augmented by electron microscopy. For this purpose lacquer replicas were made of the section from the inner zone of the flange, and additionally regions of the fracture plane were coated with carbon vapor, and the stripped films were examined. Figure 9 shows replicas of the precipitated particles in lacquer (light) and the fractured pieces extracted with lacquer (dark). The diffraction lines show thin lamellae. Electron diffraction patterns showed that these were hexagonal crystals that could consist of aluminum nitride AlN judging by their structure and lattice parameter. The particles stripped off the fracture plane showed even more clearly a form of thin, transparent platelets (Fig. 10). The diffraction patterns showed the same hexagonal structure as the particles extracted from the section plane with the lacquer.

The cause of damage therefore is the superabundant use of aluminum as deoxidizer. According to recommendations the aluminum addition was reduced by one-half. Since then no more rejects were encountered due to insufficient tensile and bend values.

In cases where aluminum and nitrogen concentrations are not too high conchoidal fractures can be eliminated by solutioning and diffusion annealing at high temperature. The annealing test with specimens of the examined flange disk confirmed (Fig. 11) that the tendency to chonchoidal fracture clearly diminishes with rising annealing temperature and time, but even a twenty-four hour annealing at 1200° C was insufficient at the high aluminum and nitrogen concentrations described to eliminate the fracture phenomena.

Cast Ingot Cracked During Forging

Friedrich Karl Naumann and Ferdinand Spies

Max-Planck-Institut für Eisenforschung
Düsseldorf

An octagonal steel ingot weighing 13 tons made of manganese-molybdenum steel containing 0.39 % C, 0.41 % Si, 1.38 % Mn, 0.014 % P, 0.009 % S and 0.32 % Mo, developed gaping cross-cracks on all eight sides in the forging press during initial pressure application. It was reported that the steel had been melted in a basic 12-ton arc furnace, oxygenated, furnished with 42 kg of 75 % ferrosilicon and 12 kg aluminum additions, alloyed with 160 kg of 80 % ferromanganese, and finally deoxidized in the ladle with 42 kg calcium silicon. The iron mold had been preheated to 125 °C and its inner surface was dusted with aluminum powder.

The block was 1780 mm long and had a diameter of 1020 mm at the top and 920 mm at the bottom. It was stripped after 10³/₄ hr, transferred into the forging furnace at 1080 °C and then kept at 1200 °C for 10 hours before forging.

The block is shown in Fig. 1. A particularly deep crack appeared in the transition to the feeder head (not visible in Fig. 1). The crystallites with conchoidal grain boundary fracture which are shown in Fig. 2 could be broken out of this crack.

For metallographic examination a plate approximately 100 mm thick was cut parallel to one of the eight planes. Its upper quarter is illustrated in Fig. 3 as seen from the lateral plane of the block, while Fig. 4 shows the etched section plane on the opposite inner side of the same section. It is apparent that the cracks have penetrated deeply into the block. They run along the primary grain boundaries. After breaking open the fracture planes showed a predominantly conchoidal structure (Fig. 5). The smooth, non-deformed fracture on the primary grain boundaries can be particularly well distinguished from the ragged ductile fracture (Fig. 6) under the scanning electron microscope.

Fig. 1. View of the forged 13-ton block

Fig. 2. Crystallites removed from gaping crack. 1 x

Fig. 3. Octagonal plane in upper quarter of block. Approx. $^1/_3$ x

Fig. 4. Longitudinal section at 100 mm depth parallel to octagonal plane. Etch: Copper ammonium chloride solution according to Heyn. 1 x

Fig. 5. Fracture specimen. 1 x

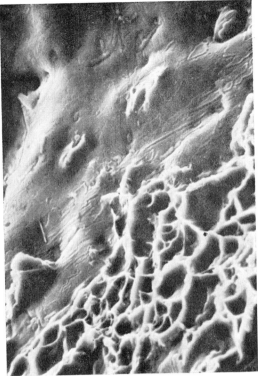

Fig. 6. Fracture plane in scanning electron microscope. Lower right: Ductile fracture, top: Conchoidal fracture. 2000 x

During microscopic examination of microsections, a coarse-meshed net of fine precipitates was found (Fig. 7). They were in part transparent and were illuminated four times while turning 360° in polarized light between crossed Nicols. Thus they were anisotropic, i. e. not glassy or of cubic structure (Fig. 8).

The melting of the steel in the basic electric arc furnace with a high content of dissociated nitrogen permitted the assumption that the precipitates were nitrides. This had been found in several prior cases of conchoidal fractures [1]. Analysis showed that the steel contained 0.022 % Al and 0.017 % N.

Electron microprobe tests showed that the precipitates were not enriched by either iron, silicon, manganese, oxygen or sulfur. In contrast to this, the aluminum radiation showed such a high intensity that this metal

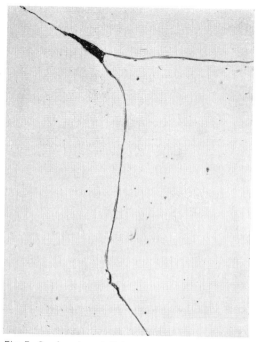

Fig. 7. Crack and precipitates at primary grain boundaries. Unetched longitudinal section. 100 x

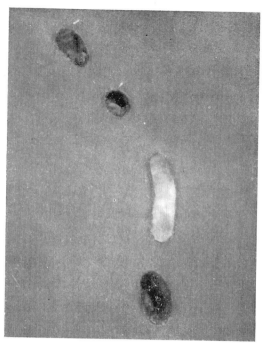

Fig. 8. Grain boundary precipitates in polarized light. Longitudinal section. 1200 x

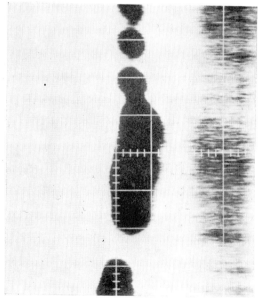

Fig. 9 a. Electron diffraction pattern

Figs. 9 a and b. Examination by electron microprobe

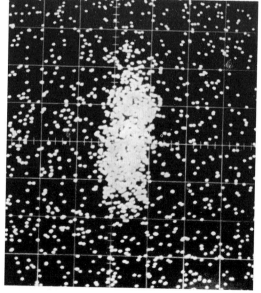

Fig. 9 b. Aluminium K_α - radiation

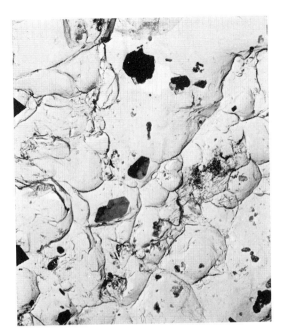

Fig. 10. Precipitates in carbon film of fracture. 5000 x

Fig. 11. Electron diffraction pattern of precipitated particle

had to be the principal constituent of the precipitates (Figs. 9 a and b). Calcium also was present. Therefore the precipitates could not consist of oxides, silicates or sulfides.

Platelet-like particles could be discerned on the conchoidal fracture planes with the scanning electron microscope (Compare Fig. 6). They were isolated by vaporizing with carbon and stripping of the film (Fig. 10). Their lattice structure was examined by electron diffraction (Fig. 11). The precipitates proved to be thin and partially transparent platelets of a hexagonal crystal lattice whose parameters resemble those of AlN.

It has been known from previous experience that such precipitates lower the cold strength and toughness of steel castings and lead to a ripping open of the grain boundaries during mechanical stress application or hydrogen loading [1]. During forging they apparently also cause hot tears according to

the determinations made here. If the data given by the manufacturer are correct it is noteworthy that the precipitates were at least in part still undissolved in spite of the long holding period at high initial forging temperature.

It may be added here that another block that was melted under the very same conditions and immediately after the defective one, was forged into a gear ring without any trouble. This ring was free of grain boundary precipitates of the kind described, as could be proved by metallographic examination. But it contained only 0.012 % Al and 0.0102 % N. A favorable circumstance may also have been that the ring was removed from the casting mold after 5¼ hr and transferred into the forging furnace at a correspondingly higher temperature. At this temperature the nitrides were perhaps not yet precipitated.

[1] F. K. NAUMANN, E. HENGLER, Stahl u. Eisen 82 (1962) 612/621

Primary Grain Boundary Cracks in Cast Ingots and Flaky Crankshafts

Friedrich Karl Naumann and Ferdinand Spies
Max-Planck-Institut für Eisenforschung
Düsseldorf

Octagonal cast ingots weighing 6.5 tons and made of unalloyed heat treated steel CK 45 according to DIN 17200, and crankshafts forged from these ingots showed internal separations during ultrasonic testing. In order to determine the cause of defect, an ingot slice and a crank arm were examined metallographically.

Figure 1 shows a section of the ingot slice after primary etching with copper ammonium chloride solution (according to Heyn). The slice was free of unusual segregations which could be confirmed by sulfur replicas according to Baumann. But it did show large numbers of fine cracks. A piece of the slice was heat treated and fractured. The cracks appeared as bright spots in the fibrous fracture (Fig. 2). Microscopic observation showed that the separations were short cracks that ran along the primary grain boundaries marked by ferrite and sulfide segregations

Fig. 1. Section of slice of a cast ingot, etch: Copper ammonium chloride solution according to Heyn. 1 x

Fig. 2. Fracture of ingot slice, heat treated. 10 x

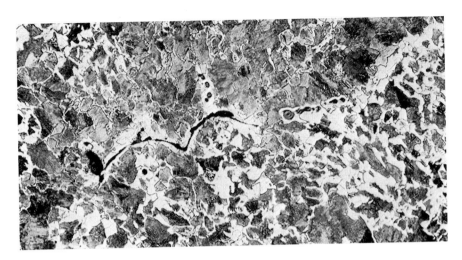

Fig. 3. Primary grain boundary separation in ingot slice, etch: Picral. 100 x

(Fig. 3). The formation of such cracks is favored by segregations of various kinds [1] [2] [3]. Their origin is not known for certain, but there are reasons to believe that the absorption and precipitation of hydrogen may be a contributing factor.

Figure 4 shows a surface view of the crank shaft arm after magnetic particle testing for cracks, while Fig. 5 shows the same section after etching with copper ammonium chloride solution according to Heyn. Both show a plurality of short cracks. But apparently a

Fig. 4. Section of surface of a crank arm after magnetic particle testing. ²/₃ x

Fig. 5. Same as Fig. 4, etch: Copper ammonium chloride solution according to Heyn. $^2/_3$ x

different kind of separation is at work in this case. The separations are predominantly transcrystalline. In the fracture of a heat treated specimen they show characteristic

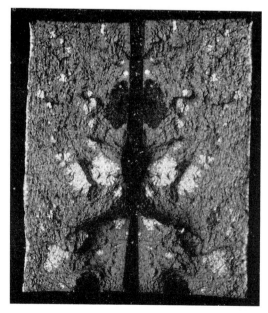

Fig. 6. Fracture of a specimen from crank arm, heat treated. 1 x
Flaky edges of specimen oxidized during heat treatment.

flakes (Fig. 6) which are small internal stress cracks whose formation is caused by the precipitation of hydrogen [4]).

This then is a case where flaky forgings were made from cast ingots with primary grain boundary cracks. This parallelity supports the often expressed opinion [1]) [5]) [6]) that both occurrences have the same origin, i. e. that hydrogen precipitation is the driving force in the formation of primary grain boundary cracks in cast ingots.

References

[1]) F. K. NAUMANN, E. HENGLER, Stahl u. Eisen 82 (1962) 612/621

[2]) F. K. NAUMANN, Archiv Eisenhüttenwes. 35 (1964) 1009/1010

[3]) Vgl. F. K. NAUMANN, FERDINAND SPIES, Prakt. Metallogr. 9 (1972) 658/665; 10 (1973) 704/707; 708/710; 711/716

[4]) Vgl. F. K. NAUMANN, FERDINAND SPIES, Prakt. Metallographie 4 (1967) 541/545 (u. 2 weitere Stellen demnächst)

[5]) ED. HOUDREMONT, Handbuch der Sonderstahlkunde, 3. Aufl. (1956) 859/861

[6]) ZD. EMINGER, Neue Hütte 4 (1959) 596/608

Cast Steel Pinion Gear Shafts with Insufficient Elongation

Friedrich Karl Naumann and Ferdinand Spies

Max-Planck-Institut für Eisenforschung
Düsseldorf

The specified elongation of 10 % could not be achieved in several hollow pinion gear shafts made of cast chromium-molybdenum steel GS 35 CrMo 5 3 that were heat treated to a strength of 90 kp/mm². The steel was melted in a basic 3 ton arc furnace and deoxidized in the furnace and in the pan with a total of 7 kg aluminum.

Figure 1 shows the fracture of a tensile specimen with low elongation and, apparently, also with low reduction of area. In some places it is coarse grained conchoidal.

Six test bars from as many melts were examined. According to the data provided by the sender the mean composition varied little and was as follows:

C %	Si %	Mn %
0,36	0,42	0,80

P %	S %	Cr %	Mo %
0,019	0,011	1,19	0,28

They are arranged according to decreasing amounts of Al·N.

The bars were normalized. In this state they showed a normal transcrystalline fracture topography. Because prior experience has shown that in the case of grain boundary precipitates the fracture follows these grain boundaries at normal temperature only if a

Fig. 1. Fracture of a tensile specimen with partially conchoidal structure. 2 x

Because melting in the basic arc furnace and strong deoxidation with aluminum gave rise to the conclusion that the damage was caused by precipitation of aluminum nitride onto the primary and austenitic grain boundaries, subsequent determinations were made of aluminum and nitrogen content and the following values were found:

high strength steel has been made tough by heat treatment, all test specimens were subjected to quenching from 850° C in water and tempering to 500° C, and then broken again. The result was the picture reproduced in Fig. 2. The fracture of bars 1 and 2 with high aluminum content is almost completely

Leiste/Specimen	1	2	3	4	5	6
Al %	0,108	0,081	0,064	0,040	0,045	0,037
N %	0,015	0,014	0,015	0,016	0,013	0,012
Al·N·10⁵	162	113	96	64	59	44

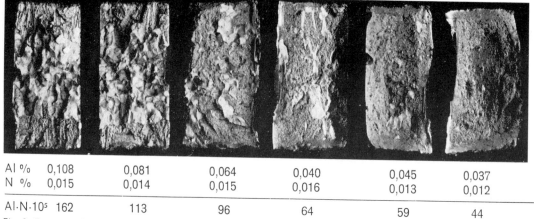

Al %	0,108	0,081	0,064	0,040	0,045	0,037
N %	0,015	0,014	0,015	0,016	0,013	0,012
$Al·N·10^5$	162	113	96	64	59	44

Fig. 2. Fracture of test specimens after heat treatment. approx. 1 x

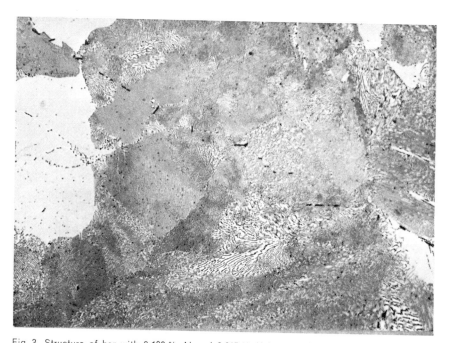

Fig. 3. Structure of bar with 0.108 % Al and 0.015 % N in normalized state as received. Etch: Picral. 500 x

conchoidal. With decreasing aluminum content the conchoidal fracture portion decreases visibly and the portion of fibrous heat treated fracture increases correspondingly. The fracture is completely fibrous at the lowest aluminum- and nitrogen content.

Rodlike or platelike precipitates were found (Fig. 3) during metallographic examination of the normalized specimens with high aluminum and nitrogen content. They were arranged in coarse-meshed networks. At another occasion these particles could be identified by electron diffraction as aluminum nitride.

Therefore, the previous assumption derived from the statements by the sender, that the exceptionally low elongation of the cast specimens was due to excessive deoxidation by aluminum, was verified.

Broken Back up Rolls from a Broad Strip Mill

Friedrich Karl Naumann and Ferdinand Spies
Max-Planck-Institut für Eisenforschung
Düsseldorf

Several back up rolls of 1400 mm barrel diameter from a broad strip mill broke after a relatively short operating time as a result of bending stresses when the rolls were dismantled. The fracture occurred in the conical region of the neck at about 600 mm diameter. The rolls were shaped steel castings with 0.8 to 1.0 % C, ca. 1 % Mn, ca. 1 % Cr, ca. 0.5 % Mo and ca. 0.4 % Ni and were heat treated to a tensile strength of 950 N/mm². Since the bending stress on mounting was only 42 N/mm² in the fracture cross section, it was evident at the outset that material defects had promoted the fracture.

Figure 1 shows one of the rolls with the broken off neck. In the centre of the fracture is a darkly tinted excentric flaw (Fig. 2). Such internal cracks occur when a cast or forged work piece is heated up too quickly and insufficiently thoroughly for the anneal. The thermal expansion of the outer zone puts the central core under tensile stress. In this case the casting has in addition obviously been warmed up from one side only.

In the case of this roll and the other broken rolls, the cracking and fracture were promoted by various casting defects. In Fig. 2 or better in Fig. 3, which shows the section of a fracture in another roll, it can be seen that the structure of the fracture was scaly apart from a narrow fine grained outer zone. Scaly fracture of this type occurs if the primary grain boundaries are weakened by the precipitation of poorly soluble nitrides, carbides or sulphides [1] [2]. It has a matt grey appearance. In the present case, several

Fig. 1. Insertion neck of roll with broken end

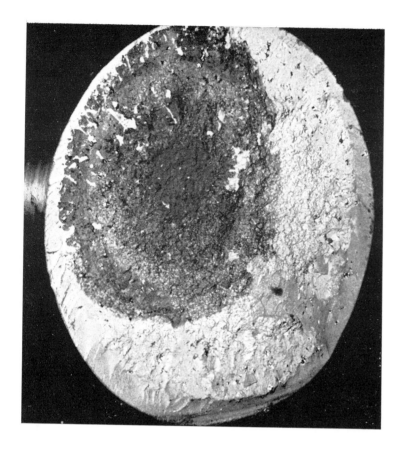

Fig. 2. Fracture surface. Ca. 0.2 x

Fig. 3. Fracture surface on outer edge of roll. 1 x

places stand out from the scaly fracture surface on account of their shiny appearance. These are open fractures, so-called primary grain boundary cracks [3]. The mechanism by which they arise is not definitely known. It is probable that hydrogen is involved in their formation as in that of flakes [4]. Since, unlike flakes, they generally occur in cast ingots or shaped castings they are called cast state flakes. Here again, however, the grain boundaries must be occupied by precipitates which make possible the precipitation of hydrogen and facilitate separation. In the core of the roll the fracture microstructure was dendritic (Fig. 4); a shrinkage cavity must therefore be present.

This was confirmed by a section next to and parallel to the fracture as shown in Fig. 5. It was subsequently ascertained that the fractures in all the rolls were at the head of the casting. Apart from the central top shrinkage cavity there were numerous microcavities which are unavoidable in such a large casting and two top crusts or double skins mark-

ed A and B in Fig. 5. As described earlier [5], top crusts are pieces of material which solidified on casting and either sank or overflowed. They differ from their surroundings only in their microstructure. The section with top crust A is shown in Fig. 6 in its natural size. Several primary grain boundary cracks are also visible.

Figure 7 shows the fine structure of one of the micro-shrinkage cavities. The last interdendritic spaces to solidify are enriched with carbon which has precipitated as lede-buritic eutectic. Carbide precipitates on the primary grain boundaries (Fig. 8) are probably also responsible for the scaly fracture and the formation of primary grain boundary cracks.

In several of the rolls the barrel edges had disintegrated over part of the circumference. These fractures also originated within the roll and run in one direction to the end face of the barrel towards the neck and in the other towards the barrel surface which they

Fig. 4. Fracture surface in core zone of roll. 1 x

A ↓

B
→

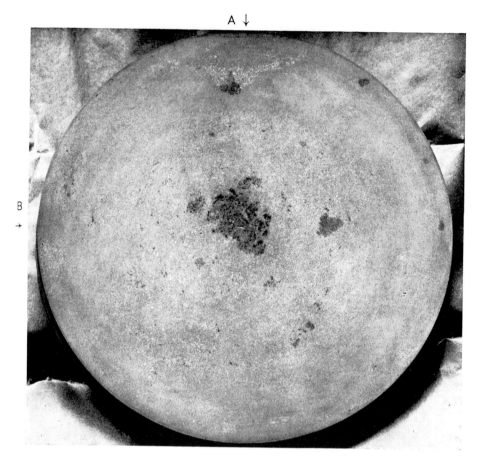

Fig. 5. Cross section through neck parallel to fracture, Heyn etch. Top crusts at A and B. 0.17 x

reach at about 150 mm from the edge. Figure 9 illustrates such a fracture. The fracture has run from right to left according to the fibre. On the right of the picture there is another fracture L — — L which has proceeded in the direction of the longitudinal roll axis. Numerous primary grain boundary cracks have been exposed by the fracture. This can be seen even better in Fig. 10 which is a section of Fig. 9 in its natural size. Irregularities in the longitudinal fracture L — — L indicate the existence of a large flaw (Fig. 11). Considerable phosphorus and sulphur segregations in a zone permeated with cavities were detected in a polished section S — — S parallel to the fracture

L — — L by primary etching (Fig. 12) and a sulphur print (Fig. 13). A top crust is clearly involved here. It must have solidified on the rising steel surface during casting and have become caught in the inward growing edge crystals.

The investigation of the rolls has thus clearly shown that both the breaking off of the neck and the disintegration of the barrel edges is caused by material defects, more exactly casting defects. The fractures on the other rolls examined were so badly rusted or contaminated that they were incapable of yielding any information.

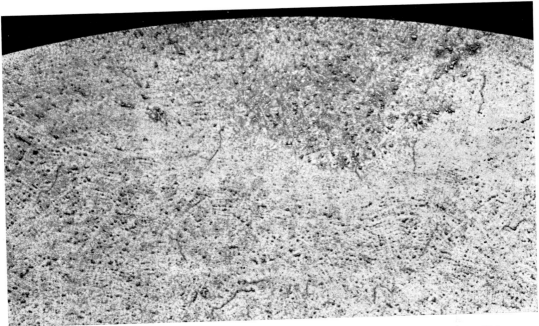

Fig. 6. Section of cross section in Fig. 5. Top crust B and primary grain boundary cracks. Etched in dilute aqueous sulphuric acid. 1 x

Fig. 7. Micro-shrinkage cavity with carbon segregation (ledeburite). Etched in picral. 50 x

Fig. 8. Primary grain boundary with precipitates of ledeburite and secondary cementite. Etched in picral. 200 x

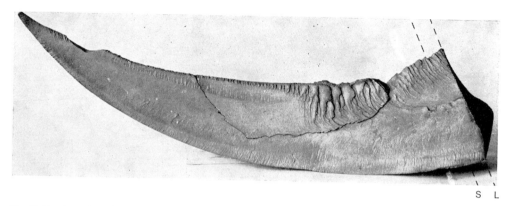

S   L

Fig. 9. Fracture piece from barrel surface. L – – L is a longitudinal fracture, S – – S a section parallel to it. 0.6 x

Fig. 10. Section from Fig. 9. 1 x

Literatur / References

1) F. K. NAUMANN, E. HENGLER, Stahl· u. Eisen 82 (1962) 612/621

2) F. K. NAUMANN, Arch. Eisenhüttenwes. 35 (1964) 1009/1010

3) E. HOUDREMONT, Handbuch der Sonderstahlkunde, 3. Aufl. unter Mitarb. von H.-J. Wiester. Berlin/Göttingen/Heidelberg (1956) 222/224

4) Zd. EMINGER, Neue Hütte 4 (1959) 596/608

5) F. K. NAUMANN, F. SPIES, Prakt. Metallographie 4 (1967) 371/

Fig. 11. Fracture

Fig. 12. Heyn etch

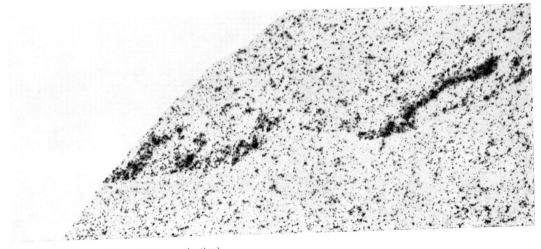

Fig. 13. Baumann sulphur print (reproduction)

Figs. 11 to 13. Section S – – S in Fig. 9. 0.5 ×

An Example of Incidence of Non-Metallic Inclusions in an Alloy Steel

P. K. Chatterjee

Ministry of Defense, Controllerate of Inspection (Metals)
Ichapur, West Bengal, India

There was once a large incidence of surface defects on the crank pins and journals and other areas of crank shafts of a high power automotive engine. The steel used was a chromium-molybdenum type of nitriding steel to the following nominal composition in wt.%:

Carbon	0.2–0.3
Silicon	0.10–0.35
Manganese	0.4–0.65
Nickel	0.4 maximum
Chromium	2.9–3.5
Molybdenum	0.4–0.7
Sulphur	0.05 maximum
Phosphorus	0.05 maximum

The defects which were initially thought to be cracks were first noticed in visual inspection after rough machining of the hardened and tempered forgings. Magnaflux test confirmed the defects but did not dispel the doubt as to whether the affected areas were open cracks or non-metallic inclusions. The defects were not in one continuous line. They were of varying lengths from 3 mm upto about 25 mm and occasionally longer. Also the defects were in different locations at different depths after turning down the diameter of the crank pin. The inspection standards allow a certain amount of latitude for non-metallic inclusions on the surface whereas open cracks can in no case be permitted. The following tests/observations were recorded on representative samples of defective crank shafts.

Examination of defective portions at low magnification

Fig. 1 shows the defective portion after Magnaflux test. The discontinuity is clearly visible.

Fig. 1. Discontinuity on the surface observed in Magnaflux test. 6×

Fig. 2 shows another area after deep etching in 20 % boiling HCl. The area shown in Fig. 2 is preferentially attacked leaving a small depression. Rugged outlines of this depression suggested that they were due to entrapping of non-metallic inclusions which have been attacked by the acid.

Examination at higher magnification

In view of the above observations, it was felt that the areas should be seen at higher magnification in unetched condition. A transverse section through defective areas polished as for microscopic examination, showed the extent and nature of distribution of the foreign material. Typical areas are shown in Figs. 3 and 4.

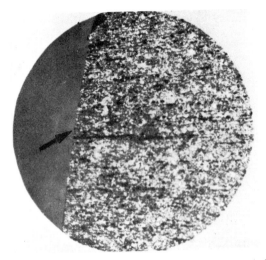

Fig. 2. Discontinuity in another area opened up after boiling in 20 % HCl. 6 ×

The unetched specimens showed that the defects as "chain and dot", the dots being globular. This observation made it clear that the defects were not cracks but entrapment of foreign bodies. To find the nature of the foreign body a specimen was subjected to electronprobe microanalysis. The result was as given below:

"The elements found to be present in the inclusion are: Mn, Si, Cr and S. No Fe was found in the inclusion. The distributions of S and Si were not uniform along the length of the inclusion. Mn content was found to be very high in the inclusion, much in excess of the matrix."

(Continued on the next page)

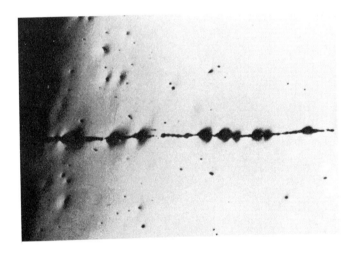

Fig. 3. Affected area in polished and unetched condition. 50 ×

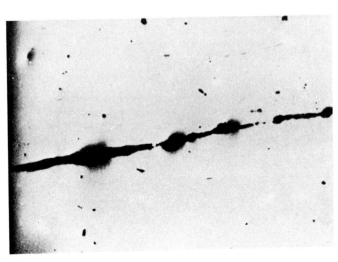

Fig. 4. Same area as in Fig. 3. 100 ×

Discussion

The simple metallographic observations conclusively proved that the defective areas were entrapment of foreign bodies, resulting from steel making/deoxidising/teeming stages. The occasionally globular nature of the foreign particles suggest that these were formed at the liquid condition of the steel. The ratio of Mn-Si as seen on electron probe microanalysis also suggests that the globules high in Mn content might have resulted in deoxidising stage, particularly the absence of Fe in some areas in the inclusion are indicative of precipitation deoxidation by Ferro Manganese/Ferro Silicon. The defects did not apparently have time to coalesce and rise up to the top.

The interesting metallographic observation was the ascent/migration of the deoxidation products as individual globules or cluster as shown in Figs. 3 and 4.

Steel Socket Pipe Conduit Cracked Next to Weld Seam

Friedrich Karl Naumann and Ferdinand Spies

Max-Planck-Institut für Eisenforschung

Düsseldorf

A steel socket pipe conduit NW 150 cracked open during pressure testing next to the weld seam almost along the entire circumference. The crack occurred in part in the penetration notch and in part immediately adjacent to it (Fig. 1).

A longitudinal section was made for metallographic examination at the still adhering portion. It could be seen immediately after etching with copper ammonium chloride solution according to the Heyn method, that the weld-jointed conduit pipes consisted of different materials (Fig. 2). While the uncracked pipe showed the light etch shading of a low-carbon steel in which the zone heated during welding was delineated only slightly next to the seam, the other pipe was etched much darker, i. e. higher in carbon, and the heated zone appeared to stand

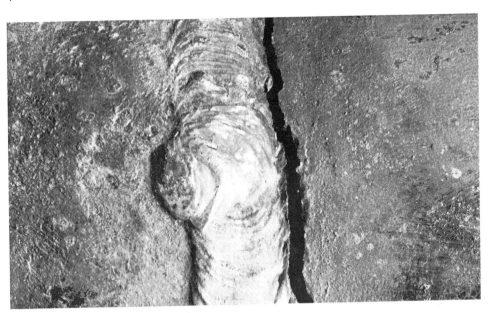

Fig. 1. View of crack region. approx. 1 x

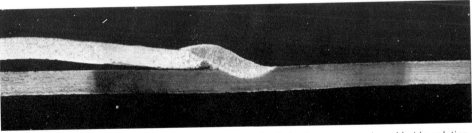

Fig. 2. Longitudinal section through weld opposite crack. Etch: Copper ammonium chloride solution according to Heyn. 1 x

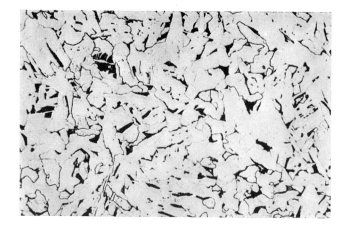

Fig. 3. Uncracked pipe, vicinity of seam

Fig. 4. Cracked pipe, unchanged structure

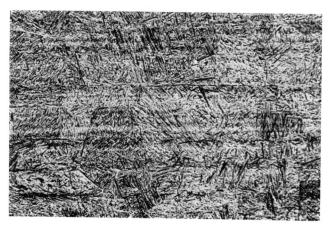

Fig. 5. Cracked pipe, vicinity of seam

Figs. 3 to 5. Structure of pipes. Longitudinal section, etch: Nital. 100 x

out darkly against the basic material. The overlapping weld was defect-free and dense.

The uncracked pipe showed a ferritic-pearlitic structure in micro-section. Except for grain coarsening and a partially spear-like formation of ferrite (Widmannstätten struc-

ture), no major changes had occurred in the vicinity of the weld seam due to the heat of welding (Fig. 3). But the cracked pipe showed a ferrite-free mixed structure in the unaffected part consisting of pearlite and bainite phases (Fig. 4). In the vicinity of the weld seam it had become coarse grained and was transformed into martensite (Fig. 5). The cracking open of this pipe next to the weld seam may be explained by the high stresses caused by the austenitic-martensitic transformation.

From the basic structure of the pipe and the delay of the transformation into the martensitic structure it had to be concluded that the pipe contained alloying elements that contributed to a rise in hardenability in addition to increased carbon content. An analysis confirmed this (Table 1). Therefore the uncracked pipe consisted of soft steel that obviously was made for this purpose, while the cracked pipe consisted of a strongly hardenable steel which contained not only more carbon and manganese than customary but also a considerable amount of chromium.

Therefore the damage was caused by a mix-up of materials that allowed an unsuitable steel to be used for the weldment.

Table 1. Chemical composition of pipes

Pipe	C %	Si %	Mn %	Cr %
uncracked	0,11	0,07	0,33	0
cracked	0,35	0,11	0,99	2,08

Poorly Drawable Steel Wire for Ball Bearings

Friedrich Karl Naumann and Ferdinand Spies
Max-Planck-Institut für Eisenforschung
Düsseldorf

A drawing plant which processed steel wire of designation 105 Cr 2 for ball bearings, had losses due to crack formation and wire breakage during drawing. In order to establish the reason for the breakage, seven fractures were submitted for investigation with contiguous wire segments on both sides of the fracture of 300 mm each.

Longitudinal and transverse sections were taken through the fracture and at the cut ends from all wire specimens. Figures 1 and 2 reproduce a longitudinal and transverse section of a fracture location. In addition to the fracture, one side contains a number of short initial cracks. Even the small-scale reproduction reveals that the crack-region has a structure different from that of other material near the surface and the core. These regions were distributed unevenly not only over the length but also

Fig. 2. Transverse section at a fracture, etch: picral. 12 x

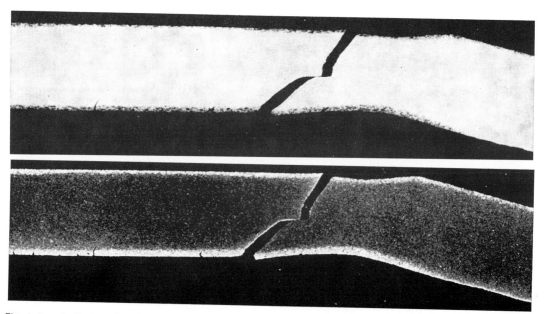

Fig. 1. Longitudinal section through a fracture, etch: picral, top: vertical illumination, bottom: oblique illumination. 4 x

Fig. 3 a. Crack region

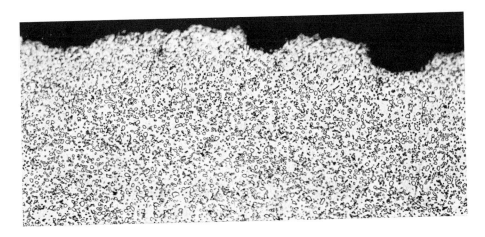

Fig. 3 b. At 300 mm distance from fracture

Figs. 3 a and b. Surface structure of the wire, longitudinal section, etch: picral. 500 x

throughout the cross section of the wire. Microscopic examination revealed that the pearlite was in lamellar form at the torn surface (Fig. 3 a), while it had coalesced into granular shape at other surface regions and in the core during the softening anneal (Fig. 3 b). The lamellar structure is less easily deformed than the granular one and therefore tears easily during drawing (Figs. 4 and 5).

Missing in the lamellar surface structure, with the exception of the remnants of a coarse network, were the pre-eutectically precipitated carbides to be expected in this steel. Some connection was suspected to exist between both phenomena; this becomes clearer in regions with higher decarburization. Such a region is reproduced in Fig. 6. Surrounding the ferritic region in the surface structure, a ring of lamellar pearlite is seen, which turns into the granular annealed structure towards the core.

The described structural phenomena were noted in all of the seven fracture regions. Their intensity always decreased with increasing distance from the fracture.

The connection between surface decarburization and the formation of lamellar pearlite during annealing, is substantiated as follows [1]: Hypereutectoid steels are annealed

Fig. 4. Structure in the cracked surface region, longitudinal section, etch: picral. 100 x

Fig. 5. As Fig. 4. 500 x

to softness most quickly and efficiently by an oscillating heat treatment about the A_1-temperature. The pearlite carbon dissolved during annealing above Ac_1, re-precipitates during slow cooling past the Ar_1 temperature by crystallizing on the undissolved carbide particles and then incorporating therein. If these particles are removed by decarburization, a eutectoid of lamellar structure is formed.

This investigation further revealed that the localized decarburization and pearlite formation was already present in the rolled wire in uneven distribution over the whole coil length.

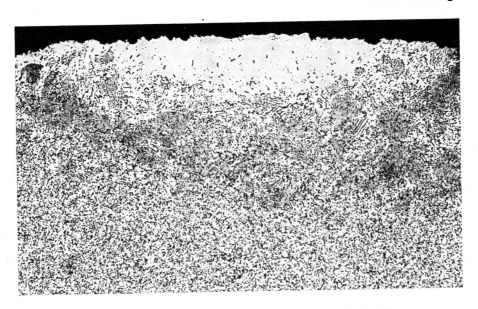

Fig. 6. Decarburized region at wire fracture, transverse section, etch: picral. 200 x

References

1) H. SCHUMANN, Metallographie, 2. Aufl., Leipzig (1958) 367

Examination of Wires for the Manufacture of Tempered Bolts

Friedrich Karl Naumann and Ferdinand Spies

Max-Planck-Institut für Eisenforschung
Düsseldorf

A bolt manufacturer complained that wires of steel 41 Cr 4 of certain shipments were prone to formation of quench cracks in the rolled threads. Since in such cases a comparison of socalled "bad" and "good" starting material (from the point of view of the user) is the surest way to establish the cause of the damage, wires were requested from shipments causing much scrap during hardening, as well as wires which were crack-free after hardening.

Thirty wire sections of equally as many rings of two "bad" shipments were examinated. In the following these are designated as S 1 and S 2. Twenty-seven sections of a "good" shipment are designated as G. Four rings of each shipment were analyzed as to their chemical composition. The analyses showed in the average the following values:

All values correspond to the specifications according to DIN 1654 and 17200. The "good" shipment can be differentiated from the two quench crack sensitive ones by a higher sulphur content which points to the smelting practice in the S. M. furnace. It also shows higher copper content which, however, should be of no significance from the point of view of crack sensitivity. But most of all higher aluminum and nitrogen contents are discernible. The wire of this shipment may therefore consist of an aluminum-killed fine-grained steel, which possesses a lower sensitivity to overheating and lower hardenability as a consequence of the presence of finely dispersed undissolved aluminum nitrides which act as deterrent to grain growth.

Cross sections were cut of all wire sections for metallographic analysis. All showed

Shipment	C %	Si %	Mn %	P %	S %	Cr %	Ni %	Cu %	Al %	N %	O %
S 1	0,39	0,28	0,73	0,017	0,006	1,10	0,08	0,03	0,006	0,002	0,007
S 2	0,38	0,27	0,72	0,017	0,006	1,10	0,10	0,03	0,006	0,001	0,007
G	0,41	0,25	0,60	0,014	0,031	1,03	0,09	0,14	0,017	0,006	0,004

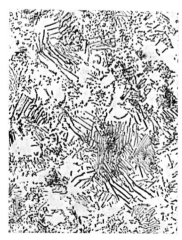

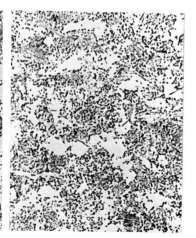

Fig. 1. S 1   Fig. 2. S 2   Fig. 3. G

Figs. 1 to 3. Core structure of the wires

annealed structures. In the structure of the wires of shipment G the cementite was somewhat stronger spheroidized (Fig. 3) than in that of shipments S1 (Fig. 1) and S2 (Fig. 2), whose lamellar pearlite structure was still recognizable. This is most likely connected with the fine grained steel's lower supercooling propensity of the austenitic transformation. All samples were case decarburized. The average depths of decarburization of all 30 samples were measured to be:

strongly and deeply under the same conditions as those of shipments S1 and S2 (Figs. 8 to 10). The tendency to stronger decarburization is therefore caused already by the melting and deoxidation process. The case structure as well as the core structure of the wires of shipment G was finer grained after decarburization annealing than those of shipments S1 and S2 (Figs. 11 to 13).

Hardness tests at temperatures of 775 to 900° C confirmed that the crack sensitive

Shipment	S1	S2	G
Depth of fully decarburized zone [mm]	0,05	0,05	0
Total decarburized depth [mm]	0,14	0,12	0,05

The wires of shipment G which were not crack sensitive are accordingly distinctly less decarburized and show no fully decarburized layer. Figures 4 to 6 display this condition.

The frequency curves of Fig. 7 express this even more clearly. Decarburization tests of 7 h at 1000° C in air showed that the wires of shipment G were decarburized only half as

steel of shipments S1 and S2 was more sensitive to overheating than the fine grained steel of shipment G. While in the specimen of S1 and S2 symptoms of overheating appeared in the fracture already after quenching from 850° C, the steel of shipment G was still fine grained after hardening from 900° C.

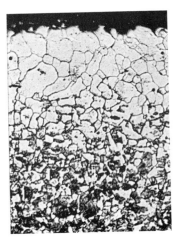

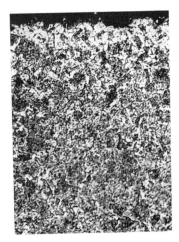

Fig. 4. S1   Fig. 5. S2   Fig. 6. G

Figs. 4 to 6. Edge structure of the wires

Figs. 1 to 6. Cross sections, etchant: picral. 200 x

End quench tests of 2 specimens of 5 mm diameter and 65 mm length were made for the purpose of comparing the hardenability A small platelet of 25 mm diameter was welded onto the end faces of the samples for protection of the bar surfaces against water spray *). Hardening temperature was 840 ° C. The results are summarized in Fig. 14. They confirm the previously reached conclusion that the steel of the crack sensitive shipments is more hardenable than that of the good shipment. The slightly higher manganese content of the shipments subject to complaints may have contributed to this, but certainly did not play the deciding part.

According to the test results the tendency of the bolts of shipments S 1 and S 2 to crack in the thread during hardening is in all probability primarily due to their great sensitivity to overheating and to their hardenability caused by their method of metallurgical manufacture. A stronger decarburization of the case doubtlessly also contributed to this, which could not be prevented by working because the thread was rolled. It is well

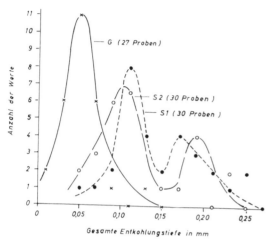

Fig. 7. Frequency curves of the decarburization depth

*) The wires were too thin for the production of standard samples of 25 mm diameter. The results of tests of non-standardized samples are comparable with each other.

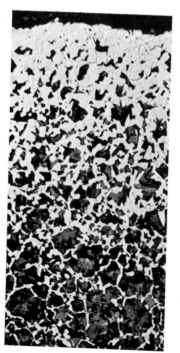

Fig. 8. S 1

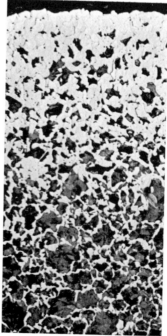

Fig. 9. S 2

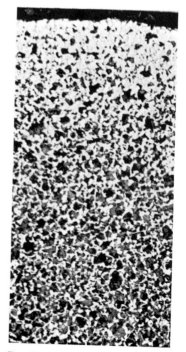

Fig. 10. G

Figs. 8 to 10. Edge structure of the wires. 50 x

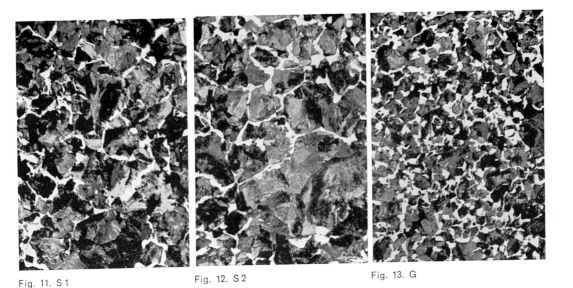

Fig. 11. S 1   Fig. 12. S 2   Fig. 13. G

Figs. 11 to 13. Core structure of the wires. 100 x

Figs. 8 to 13. Structure after 7 h decarburization anneal at 1000° C in air, cross sections, etchant: picral. 100 x

known [1]) that stresses occur during hardening of case decarburized parts through an advance of the γ-α-transformation which can lead to quench cracks. Incidentally, hardening tests by the bolt manufacturer on notched bars showed that quench cracks did not occur if the specimens were turned down before hardening or if the notch was not beaten by a chisel, but was machined.

[1]) H. BÜHLER, E. HERRMANN, Einfluß einer Randentkohlung auf die Rißgefahr bei der Härtung von unlegierten Werkzeugstählen, Stahl und Eisen 82 (1962) 622/629

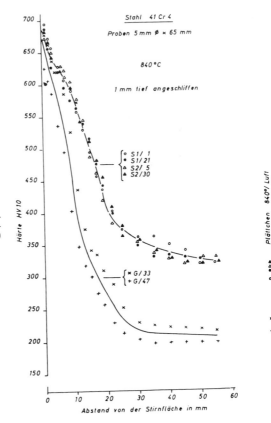

Fig. 14. Results of end quench tests

An Example of Decarburisation in Alloy Steels and Its Effect on Further Processing

P. K. Chatterjee

Ministry of Defense, Controllerate of Inspection (Metals)
Ichapur, West Bengal, India

Case history

One percent Chromium-Molybdenum steel — a low alloy constructional steel is very widely used for high tensile applications e.g. for manufacture of high tensile fasteners, heat treated shafts and axles, for automobile applications such as track pins for high duty tracked vehicles etc. The steel is fairly through hardening and heat treatment does not present any serious difficulty. It is also amenable to induction hardening and can therefore be harnessed in applications where high core strength and toughness and very high hardness on surface are required. However care is still required in processing to avoid decarburisation. In applications requiring induction hardening of the outer layers, this may present defects making the components unacceptable.

In an application of track pins for tracked vehicles, bars about 22 mm dia were required in heat treated and centreless ground condition prior to induction hardening of the surface. Indifferent results were obtained in induction hardening; cracks were noticed and patchy hardness figures were obtained on the final product in several batches. The details of chemical analysis and mechanical strength on the bulk as obtained, are given below:

Chemical composition (wt.-%)
Carbon 0.40
Manganese 0.73
Chromium 1.15
Molybdenum 0.28

Mechanical properties
(on heat treated bars intended for induction hardening)
Yield .... 70 kp/mm^2
Ultimate tensile strength .... 88 kp/mm^2
Elongation .... 18.0 %
Impact .... 45.52 ft lbs (Izod)

The stock when examined by magnaflux method revealed presence of longitudinal cracks of varying lengths, the longer ones being 100 mm in length and smaller ones about 10 to 20 mm. The cracks were not continuous but in the same line. It was apparent the defective areas were continuous but in some areas the cracks were removed in centreless grinding, which might be due to the greater depth of the defect in some regions or greater depth of the material having been removed in machining or in centreless grinding. The observations of the magnetic crack detection test were confirmed by deep etch test in 10% hydrochloric acid.

Metallographic examination of transverse sections through the defective areas showed decarburisation to varying degrees i. e. from partial to total decarburisation. Photomicrographs in Fig. 1 and 2 show partial and complete decarburisation respectively.

Conclusion

The observations suggest the defects to have originated at the stages of ingot making and rolling. The surface defect with shorter depth and relatively sharp outline (Fig. 1) is apparently a crack from original ingot/bloom stage gradually enlarged in rolling/heat treatment. The defect in Fig. 2 with irregular contour and greater width is suggestive of entrapment of slag/oxides inclusion. The widening of the outline might have been caused by oxidation of an initially present fissure which was filled up by low melting oxides (clinkers) in the furnace bed. This is apparently the reason for

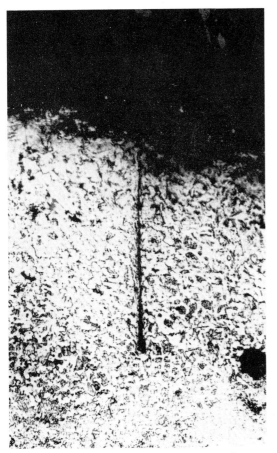

Fig. 1. Crack through partially decarburised area. 150 ×

Fig. 2. Slag entrapment in totally decarburised area. 150 ×

complete decarburisation of the area with original surface defect which opened up further in the oxidising atmosphere of the furnace with low melting clinkers from scale and furnace lining filling up the crevice of the original defect.

The microscopic examination furnished the clue and solution to problem of patchy hard-ness in induction hardening due to decarburisation of the outer surfaces, and shows the importance of this simple yet powerful tool of the metallurgical microscope in the hands of a Works Metallurgist. The photomicrographs are good examples of partial and total decarburisation not easy to come by in alloy steels containing carbide forming elements.

Fractured Suspension Bar

Friedrich Karl Naumann and Ferdinand Spies

Max-Planck-Institut für Eisenforschung
Düsseldorf

In a concrete structure a ceiling was hung on flat bars of 30 x 80 mm² cross section. The bars were borne by a slit steel plate and supported by tabs that were welded onto the flat sides (Fig. 1). One of the bars fractured during mounting when it was dropped from a height of about 1 m onto the opposite support.

The fracture was a grainy forced rupture as can be seen from Fig. 2. It propagated from one of the fillet welds. For metallographic examination, longitudinal and transverse sections were made in the fracture zone but at some distance from the fracture itself, i. e. outside of the region of weld effects. Macro-etchings and sulfur prints according to Baumann showed the existence of distinct grain segregation (Fig. 3). Therefore the material was unkilled cast steel subject to aging. Etchings according to Fry proved through "force action figures" that the bar indeed was aged not only in the area of the weld (Fig. 4 a), but also, and especially in places where the welding stresses and heat were not effective (Fig. 4 b). The cold deformation which is particularly evident in the figures that were more strongly attacked may have occurred during straightening of the bars. Otherwise the steel was free of defects. The structure was finegrained up to the immediate area of the weld.

Chemical analysis showed the following composition of the steel:

C %	Si %	Mn %
0,07	< 0,01	0,43

P %	S %	N %
0,006	0,023	0,004

Fig. 1. Side view of upper end of suspension bar with welded-on tabs. approx. ¾ x

Fig. 2. Fracture view, arrow = crack origin approx. ¾ x

Fig. 3. Baumann print of transverse section. approx. ¾ x

The quantities of elements that could have an effect on suitability of welding, such as carbon, phosphorus, sulfur and nitrogen, are decidedly low. But DIN 17 100 emphatically points out that killed steels are preferable to unkilled ones in welding constructions.

Tensile tests on 20 x 30 x 140 mm bars machined longitudinally showed the values reproduced in Table 1. They corresponded to the standard for U St 34 according to DIN 17 100.

Figure 5 shows notch impact toughness as a function of testing temperature of ISO-V-

Fig. 4 a. Area of fracture, arrow = crack origin

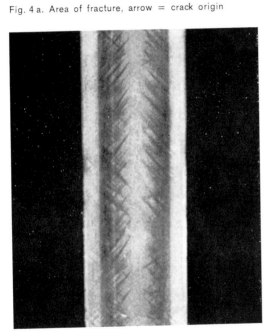

Fig. 4 b. Approximate distance of 200 mm from fracture

Figs. 4 a and b. Longitudinal sections, etch according to Fry. approx. ³/₄ x

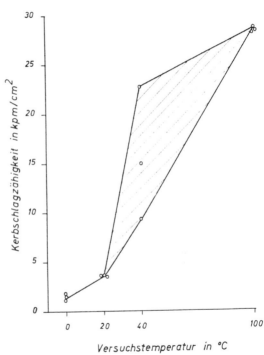

Fig. 5. Notch impact toughness as function of impact temperature

Table 1. Results of tensile tests

Yield Point		Tensile Strength	Elongation $L_0 = 5d$	Reduction in Area
upper kp/mm²	lower kp/mm²	kp/mm²	%	%
21,8	21,7	35,3	35,5	70
23,7	23,0	35,5	38,4	70

notch specimens. The values at $+40°\,C$ lie already in the scatter region and are lowest at room temperature.

Thus a steel was selected for this important construction that was prone to aging and that in fact had aged through cold deformation during straightening and then was welded yet. The bar could withstand mounting and subsequent static loading as long as it was treated with care, as could be expected from the good deformation characteristics of the static tensile test. The question is, however, whether occasional impacts or shocks can be assuredly avoided. This risk could have been eliminated if a killed steel of quality groups 2 or 3 according to DIN 17 100 had been used.

Fractured Chain Link

Friedrich Karl Naumann and Ferdinand Spies

Max-Planck-Institut für Eisenforschung

Düsseldorf

A chain link which was part of the hoisting mechanism of a drop hammer broke after three or four months' service. It was reportedly manufactured of the heat resistant steel 30 CrMoV 9 (Material No. 1.7707).

Verification of the chemical composition yielded the following values:

The steel therefore is lacking in molybdenum; the rest of the composition approximately corresponds to the specification. The deviations are insignificant in connection with the damage.

The fracture of the chain link had a conchoidal structure and ran along the austen-

C %	Si %	Mn %	P %	S %	Cr %	Mo %	V %
0,32	0,18	0,60	0,022	0,013	2,34	0,01	0,09

Fig. 1. Fracture. 20 x

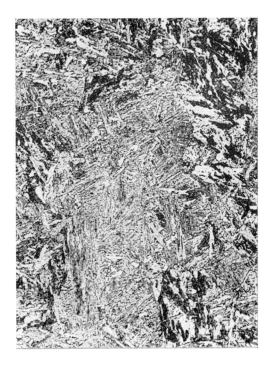

Fig. 2. Structure after heat-treatment, etchant: Picral. 100 x

itic grain boundaries (Fig. 1). Such fractures are characteristic results of strong over-heating. They occur because slightly soluble sulfidic or oxydic impurities of the steel dissolve at very high temperatures and later during cooling deposit at the austenitic grain boundaries as a submicro-scopically fine dispersion[1] [2]. These de-posits lower the strength and decrease the plasticity which is especially important in chain links. They, therefore, favor the pre-mature appearance of a fracture attribut-able to low ductility.

Figure 2 shows the coarse-grained, coarse acicular heat-treated structure of the chain link, which confirms overheating. Since temperatures in excess of 1150° C are re-quired for the solution of the forementioned impurities, it is more probable that the real damage was done during the heat-up forg-ing (drop-forging) and could not be re-moved during heat-treatment.

Literatur/References

[1] E. HOUDREMONT, Stahl u. Eisen 72 (1952) 1536/1540
[2] Vgl. a. F. K. NAUMANN, F. SPIES, Prakt. Metallo-graphie 8 (1971) 667/674

Broken Eyebolt 1¼ In.

Gisela Brunsmann and Ödön Szabó

Metall-Labor, Brown Boveri & Cie

Baden

After several years' use, an eyebolt suffered brittle fracture in the first turn of the thread. The fracture surface is shown in Fig. 1. The fracture started at the notch at the root of the thread. Neither localised material defect nor an old crack are present. The result of the chemical analysis was as follows:

C = 0.52 %; Si = 0.14 %; Mn = 0.60 %; Cr = 0.17 %; Ni = 0.10 %; P = 0.056 %; S = 0.040 %; N_{total} = 96 ppm; N_{mobile} = 60 ppm.

According to this, the eyebolt was not manufactured from the prescribed quality St 37-2N but from steel with a higher carbon content. In addition, the phosphorus and nitrogen contents were rather high.

The hardness of the defective eyebolt was HV_5 = 227 to 230 kgf/mm². Notched bar impact tests (DVM) yielded α_K = 0.7 to 1.5 kgf · m/cm².

A longitudinal section through the fracture site showed a structure of lamellar pearlite with a coarse ferrite network at the former austenite grain boundaries (Fig. 2). The fracture was fissured and made up mainly of transgranular cleavage fractures. As shown by Fig. 3, these cleavages occurred not only in the ferrite but also in the pearlite grains.

It is well known that pronounced slip planes and cleavage cracks often form in ferritic steels which have become brittle due to ageing. On the other hand, this type of fracture in a predominantly pearlitic steel seemed to us unusual. The behaviour of the material

Fig. 1. Fracture surface on the eyebolt. 2 x

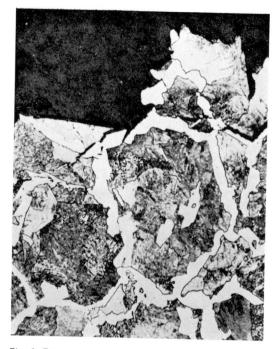

Fig. 2. Fracture and subsidiary cracks in the longitudinal section. 200 x

Fig. 3. Cleavage of a pearlite grain. 1250 x

in the vicinity of the fracture could be observed exactly by an electron microscopic examination of a carbon replica. Figure 4 shows a slip plane in the pearlite where the two crystal halves are displaced by about half the lamella spacing. If the stress was increased the pearlite grain cleaved along the slip lines shown above (Fig. 5).

In our opinion this phenomenon can be explained only by a decrease in the deformability of the α-iron matrix, i. e. by the blocking of dislocations. In normally deformable α-iron, the thin carbide lamellae of the pearlite are readily deformable.

The typical brittle cleavage fracture surface on a notched bar impact test specimen taken from the damaged eye bolt can be seen in Fig. 6.

The tendency of the steel to ageing was established by notched bar impact tests on specimens taken from the defective eye bolt and normalized at 900° C. One specimen was examined in this condition and the other after

strain ageing (10 % deformation + 1/2 h/ 250° C). The results were as follows:

	notched bar impact strength [kpm/cm²]	hardness HV [kp/mm²]
normalized specimen	3,1	223–232
aged specimen	0,8	274

There is thus a clear tendency to ageing. The microstructure of the aged notched bar impact test specimen is shown in Fig. 7.

Conclusions

The investigation showed that the eye bolt had suffered brittle fracture. Instead of the specified steel quality St 37-2 N, a steel with ca. 0.5 % C had been used.

The microstructure with the coarse ferrite network indicates that the forged eye bolt had been normalized either at too high a temperature or not at all. In any case the

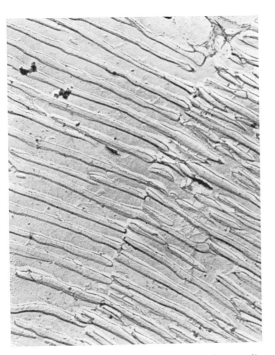

Fig. 4. Slip line in a pearlite grain. Carbon replica. 15 000 x

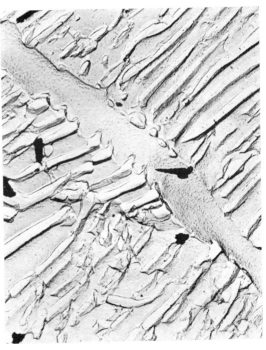

Fig. 5. Cleavage of a pearlite grain. Carbon replica. 10 000 x

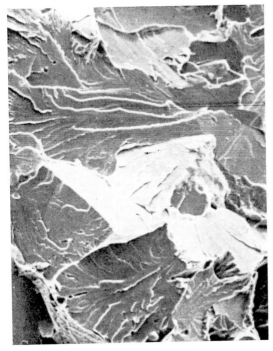

Fig. 6. Brittle cleavage fracture surface on a notched bar impact test specimen from the broken eyebolt. Scanning electron micrograph. 500 x

Fig. 7. Microstructure (with cleavage crack) of an aged notched bar impact test specimen. Normalized at 900° C, 10 % deformed and aged 1/2 h at 250° C. 500 x

anneal at 900° C produced a considerably more finely grained structure. In addition, the nature of the fracture and the results of the notched bar impact tests showed that in spite of the high C-content, the eye bolt had become brittle as a result of ageing.

Since the embrittlement of such eye bolts can lead to serious accidents, the notched bar impact strength of the eye bolts in the store and in use in the factory should be subjected to spot checks.

We would like to thank Dr. T. Geiger, Forschungslaboratorium der Fa. Gebr. Sulzer, Oberwinterthur, for the determination of the nitrogen content. The fractographic investigation was undertaken with the scanning electron microscope at the Eidgenössischen Materialprüfungsanstalt, Dübendorf.

Fracture of 99.90 Pure Tin Tubes

Karin Dieser
Hoechst AG
Frankfurt/Main-Hoechst

1. Description of damage

On account of their high resistance to corrosion, tubes made of 99.90 pure tin were to be used for packaging a chemical compound. Before filling, however, a number of the tubes cracked on bending and underwent brittle fracture whilst others remained ductile and showed no sign of failure. There was no question of the cause of failure being corrosive attack by the contents but rather the condition of the material.

2. Metallographic preparation

Longitudinal sections were cut from the externally white painted tubes with scissors and mounted in casting resin. They were wet ground on silicon carbide paper down to grade 600 and then dry on grade 600 paper which had been rubbed over with wax.

Since in view of the thinness of the specimens (0.15 mm) only poor results as regards edge and crack edge definition could be achieved with electrolytic polishing, the tin sections were prepared by mechanical methods. They were first of all polished in an automatic holder on a special cloth prepared with 1 μm diamond paste until all visible grinding marks had been removed. This was followed by hand polishing with alumina no. 1 on a felt cloth and then with alumina no. 3 on velvet. The surface deformation was removed between whiles by etching with 5 % alc. hydrochloric acid. A vibration machine proved very useful for the fine polishing although it took longer than by hand. The best edge definition was achieved using the vibration method.

All the etching experiments involved a grain surface attack and hence produced a rather strong surface relief from which the grain boundary cracks could not clearly be differentiated. The sections were therefore examined unetched in polarised light in which the grain surfaces of the tetragonal tin showed up at different brightnesses.

3. Metallographic and chemical investigations

As can be seen from Figs. 1 and 2, the microstructure of the unusable tubes differs clearly from that of the ductile tubes. The brittle tubes (Fig. 1) have much smaller grains and are cracked from the surface down along the grain boundaries. Some of the cracks in the tensile region are gaping open but there are also cracks under the paint on the outside surface of the tubes. In all the specimens the grains are polyhedral and show no signs of cold deformation because the recrystallization temperature of tin is below room temperature. No grain boundary precipitates could be detected in the microscope.

There is also a clear difference in hardness between the good and bad tube material. The microhardness HV0.025 of the good tubes is 6.7 to 7.6 and is clearly less than the values of 10 to 11.3 measured on the broken tubes.

In DIN 1704 the aluminium and zinc contents of 99.90 pure tin are limited to 0.002 wt % each. The Tin Research Institute in England has particular experience in the analysis of tin materials and, through the agency of the Tin Information Office, Düsseldorf, kindly undertook to determine these elements. Whereas zinc could not be detected in either of the tube sorts, there were clear differences in the aluminium content of the ductile and brittle specimens.

The following values were measured:

ductile tubes 0.0005 wt. %
brittle tubes 0.035 wt. %

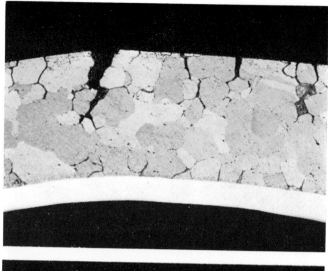

Fig. 1. Tin tube which cracked on bending (longitudinal section), unetched, polarised light. 200 x

Lackschicht/paint layer

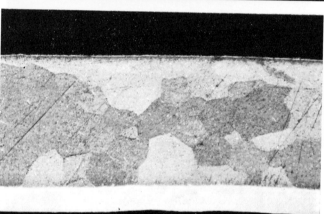

Fig. 2. Ductile tin tube (longitudinal section), unetched, polarised light. 200 x

Lackschicht/paint layer

4. Conclusions

It follows from the results of the investigation that the tendency of tin tubes to fracture is connected with the too high aluminium content of the material used. The influence of small amounts of Al up to 0.5 % on the properties of tin is described in the literature[1]).

According to this the strength and hardness are increased by only minute amounts of aluminium and then on ageing — particularly in the case of rolled material — embrittlement commences at the surface which leads to cracks or fracture on bending. This was the behaviour exhibited by our defective tubes. Their finer grain could also be attributable to contamination of the tin by aluminium. Unfortunately we could not discover from the manufacturer how the aluminium found its way into the tin.

References

[1]) HANSON, SANDFORD, J. Inst. Met. 56 (1935) 191

Section IV:
Environment-Related Failures

Hydrogen Damage/Corrosion 345-417

Stress Corrosion Cracking Failures in Components Made of Austenitic Chromium Nickel Steels

Egon Kauczor

Staatliches Materialprüfungsamt an der Fachhochschule
Hamburg

Stress corrosion cracking, also referred to as SCC, occurs mainly in metallic materials and is a form of corrosion in which cracking occurs if the following three factors occur simultaneously:

1. A material with a susceptibility of stress corrosion cracking
2. Internal or external tensile stresses
3. A corrosive medium for the particular material

The cracks that are initially very fine run either through the grains (transcrystalline) or follow the grain boundaries (intercrystalline) depending on the alloy and the corrosive medium. Usually no corrosion products can be detected macroscopically and hence SCC damage is often first seen when the stressed section, reduced in area by the numerous cracks, is no longer capable of supporting the applied stresses and fractures. The virtually undeformed cleavage fracture, which is generally free from corrosion products, is typical of stress corrosion cracking[1].

The mechanism of stress corrosion cracking has not yet been completely explained. However, in spite of the various theories, effective measures against SCC failure are possible as the nature and causes of the failure are known. The aim of this work is to show examples of SCC failures, how they occurred in practice, and to describe preventative methods based on practical experience.

The austenitic stainless steels are particularly susceptible in chloride containing solutions and in the presence of moist organic chlorine compounds at high temperatures and pressures as well as in concentrated leaching solutions[4] [7] [10]. The susceptibility to chloride increases with increasing concentration and increasing temperatures above 50°C. At high temperatures SCC can be initiated by even a minimal amount of chloride. Below 50°C the danger of SCC attack is low but it can appear after very long periods[7]. The cracks are usually very branched (fig. 1) and transcrystalline (fig. 2). Stress corrosion cracking can be inferred from the typical root-like branching which can sometimes be detected in X-ray photographs (fig. 3) or even in the form of visible cracks on the surface (figs. 4 and 5).

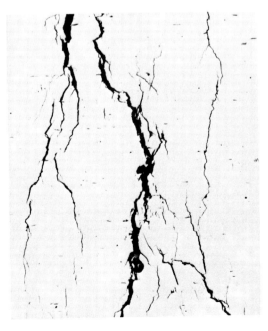

Fig. 1. Typical heavily branched stress corrosion cracks in an austenitic chromium nickel steel. Unetched micro-section. 100 ×

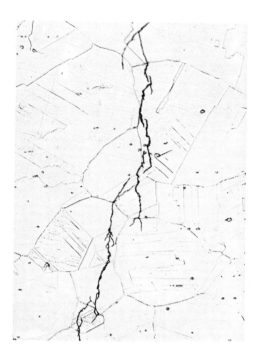

Fig. 2. Section at a higher magnification showing a propagated crack. Taken from the micro-section in fig. 1. Etched in V2A etch to show the transcrystalline crack path. 500 ×

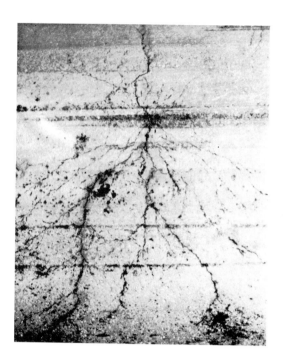

Fig. 4. Stress corrosion cracks and pitted regions on a tube of X5 CrNi 18 9. 4 ×

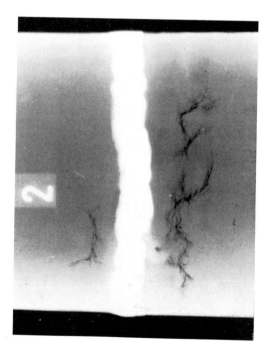

Fig. 3. X-ray photograph of stress corrosion cracks in the region of a circumferential welding joint on a tube of X10 CrNiTi 18 9. 1 ×

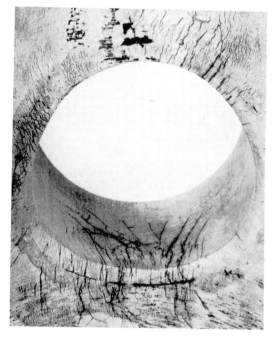

Fig. 5. Stress corrosion cracking initiated in a chloride containing medium by welding stresses in a welded support on a tube of X5 CrNiMo 18 10. 1/2 ×

Fig. 6. Unetched micro-section through two pitted regions on the wall of the tube in fig. 4. 100 ×

Fig. 7. Unetched longitudinal section from the threaded region of bolt in X10 CrNiTi 18 9 that failed by stress corrosion cracking. 8 ×

When pitting and stress corrosion cracking appear together (figs. 4 and 6) as is often observed in the molybdenum free austenitic chromium steels, the SCC cracks tend to run out from the pits. This can be explained by increased chloride concentration in the pit compared to that of the external electrolyte[5]. Fig. 6 shows stress corrosion cracks running from the pitted areas in an unetched micro-section taken from the wall of the tube in fig. 4. At low magnifications the often extremely fine branches cannot always be clearly seen. It is, therefore, obvious that cracks such as that on the left of fig. 6 should be attributed to fatigue corrosion. If all possible doubt cannot be excluded, as is the case here, by means of a macrograph (fig. 4), alternate polishing and etching (intermediary etching) should provide an extensive deburring of the cracks. The micro-section can then be evaluated at higher magnifications.

In austenitic cast steel and also in the matrix of the austenitic cast irons the SCC cracks run in straight lines and branch off at sharp angles, usually right angles. This is presumably attributable to the erratic distribution of internal stresses in the cast structure resulting from the rolling and forging textures[4]. The various types of crack behaviour can be distinguished in the fracture surfaces, as shown in the comparisons of optical micrographs of the microsection with the scanning electron micrographs of the corresponding fracture surfaces in figs. 7 to 10. The fracture surface of the bolt forged in X10 CrNiTi 18 9 (fig. 7) shows the typical feather-like structure of an SCC failure in the scanning electron micrograph[2], whereas the cleavage surfaces shown in the micrograph of a fractured austenitic cast iron (fig. 10) could almost be mistaken for a grain boundary fracture. However, the micrograph taken at a higher magnification from the microsection clearly shows the transcrystalline nature of the crack (fig. 11).

Internal tensile stresses that can initiate stress corrosion cracking may be caused by cold-working e.g. cold-rolling, straightening, edging, punching, cutting, and stamping. Critical residual stresses can also be induced by machining operations using blunt tools or clamped tools that are too long, together with excessive contact pressure[7]. In this case SCC which is limited to the effected surface of the

Fig. 8. Unetched micro-section of a cracked pump casing of austenitic spheroidal graphite cast iron. 100 ×

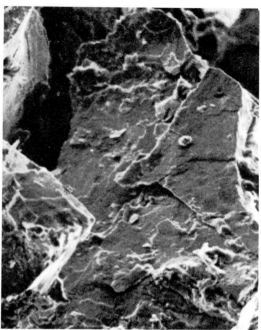

Fig. 10. Scanning electron micrograph of a fracture face from the cracked pump casing in fig. 8. 400 ×

Fig. 9. Scanning electron micrograph of a fracture face from the bolt in fig. 7. 400 ×

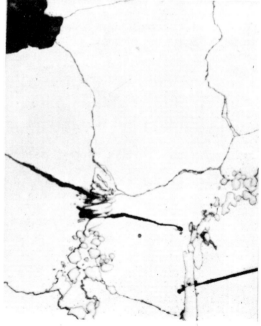

Fig. 11. Micrograph from the section of the wall of the pump casing taken at a higher magnification. Etched in V2A etch. 500 ×

otherwise stress-free work piece, gives rise to flake-like spalling[11]). Static tensile stresses can develop during welding, either from shrinkage, from subsequent dressing of the weld seam, or from slag removal by grinding (residual stresses from grinding)[6]), especially if coarse grinding wheels are used. For this reason pickling should be used to remove the slag formed on welds made without a protective atmosphere[4]). If any one of the three basic requirements for SCC can be eliminated, then it cannot occur. The methods used in its prevention stem from this fact.

All possible measures of reducing tensile stresses in the component should be taken e.g. avoidance of sharp changes of section. Also the likelihood of the medium concentrating in crevices or hot spots should be as low as possible. Residual stresses developed during production can, if the size and shape of the component are suitable, be reduced by stress relieving at temperatures of about 900°C. However, this is only possible with stabilised austenitic steels, which are not susceptible to grain boundary attack, and with the ELC grades containing a maximum carbon content of 0.03%, providing the annealing time is not too long. In cases of doubt, homogenization at a temperature of between 1050°C and 1100°C followed by accelerated cooling should be used. This also applies to those molybdenum containing chromium nickel steels which are not susceptible to grain-boundary attack in order to prevent embrittlement by precipitation of the sigma phase. The cooling following annealing has to be carefully controlled because of the high coefficient of expansion of austenitic steels to

prevent large temperature differences in the structure leading to the formation of new tensile stresses[3]). Where annealing is not possible the residual tensile stresses in the surface can be converted to compressive stresses by shot blasting. This often suffices to protect the still stressed underlying regions from stress corrosion cracking. Therefore, shot blasting is preferred if slag removal from a welded joint by pickling is not possible[4]). Since the coherent fine oxide film, necessary for the corrosion resistance of the alloy, can only form at smooth surfaces, no shot with sharp edges may be used that might give a too rough surface. From the operating point of view, care must be taken to make sure that the chloride concentration and the temperature of the medium, as well as the mechanical stressing of the component are no higher than absolutely necessary.

From the material side, the danger of stress corrosion cracking in the austenitic chromium nickel stainless steels can be reduced by replacing them with grades containing over 20% nickel. In more severe cases, the ferritic chromium steels, which are practically insensitive to SCC, can be used. However, a lower resistance to pitting corrosion has to be allowed for[9]). In really difficult cases, therefore, other alloys e.g. from the range of nickel alloys, or the possibility of using the somewhat difficult to implement method of cathodic protection[7]), have to be considered.

Acknowledgements

I am very grateful to Frau *Hannelore Schönhoff* for the preparation of the scanning electron micrographs.

References

1) DIN 50 900 Teil 1: Korrosion der Metalle, Begriffe, Beuth Verlag GmbH, Berlin und Köln (1975)

2) ENGEL-KLINGELE: Rasterelektronenmikroskopische Untersuchungen von Metallschäden, Gerling Institut für Schadenforschung und Schadenverhütung, Köln (1975)

3) T. G. GOOCH: Welding and the Corrosion Resistance of Austenitic Stainless Steels, in: The Influence of Welding and Welds on the Corrosion Behaviour of Constructionsn Public Session International Institute of Welding, Tel Aviv, Israel (1975)

4) G. HERBSLEB, Korrosionsprobleme an Schweißverbindungen hochlegierter Stähle, Schriftenreihe schweißen + schneiden, 5. Jahrgang (1976), Bericht 3

5) D. HIRSCHFELD, Erfahrungen mit Warmwasserbereitern aus nichtrostendem Stahl, in: Korrosion in Kalt- und Warmwassersystemen der Hausinstallation, Deutsche Gesellschaft für Metallkunde e.V., Oberursel (1974)

6) F. W. HIRTH, R. NAUMANN, H. SPECKHARDT, Zur Spannungsrißkorrosion austenitischer Chrom-Nickel-Stähle, Werkstoffe und Korrosion (1973) 349/355

7) MANNESMANNRÖHREN-WERKE, Lexikon der Korrosion, Band 1, Düsseldorf (1970)

8) F. K. NAUMANN, F. SPIES, Durch Korrosion undicht gewordene V2A-Rohre, Praktische Metallographie 13 (1976) 387/389

9) F. W. STRASSBURG, Schweißen nichtrostender Stähle, Fachbuchreihe Schweißtechnik Bd. 67, Deutscher Verlag für Schweißtechnik GmbH, Düsseldorf (1976)

10) VEREIN DEUTSCHER EISENHÜTTENLEUTE, Prüfung und Untersuchung der Korrosionsbeständigkeit von Stählen, Verlag Stahleisen GmbH, Düsseldorf (1973)

11) E. KAUCZOR, Spannungsrißkorrosion — Beschädigtes Stück einer Schneckenwelle, Praktische Metallographie 11 (1974) 353/356

Corrosion of a Flue Gas Inlet Foot

Fulmer Research Institute Ltd.
Buckinghamshire, England

The inlet foot (Figs. 1 and 2) was part of a fire prevention inert gas system installed in an oil tanker. The material of the foot was 2.38 mm thick Incaloy 825 plate welded using INCO welding rod No. 135. Perforation of the plate had occurred at the top and sides of the cylindrical section and near the top of the side and back walls of the foot.

Washed flue gas from the oil fired boilers enters the foot through the cylindrical inlet at $\sim$ 750°F and is further cooled by sprayed sea water to $\sim$ 600°F as the gas passes through the foot. In order to prevent the possibility of combustion of hydrocarbons the O_2 content of the gas was kept below 10% by volume; a typical composition is [in vol.%]

Figure 2 shows the top half of the foot and gives a picture of the extent and distribution of the perforations. Figures 3 and 4 are internal views of the top of the cylindrical section and the junctions between the side, back and top of the foot. They show that the internal surface is covered with corrosion product and pitted and perforated in several areas. It can also be seen that pitting is no worse at the welds than in the plate. The corrosion product was generally grey in colour but interspersed with rust and a black carbonaceous deposit.

Chemical analysis revealed that there was a factor of 10 difference in the sulphate content

CO_2	O_2	SO_2	N	H_2O + solids
12,0—14,5	2,5—4,5	0,02—0,03	77	Remainder

Fig. 1. General view of inlet foot

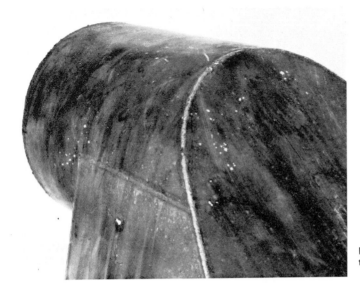

Fig. 2. Top half of inlet foot illustrating extent and distribution of perforations

Fig. 3. Inside top of inlet foot

Fig. 4. Inside junction between top, side and back of inlet foot

between the corrosion product and the carbonaceous deposit. The results were:

Corrosion product

0,60 % SO_4^{2-}

carbonaceous deposit

6,20 % SO_4^{2-}

There was also a trace of chloride in the corrosion product but neither sulphite nor sulphide was present in either sample.

Sections taken for metallographic examination were etched electrolytically in 5% H_2SO_4. Figs. 5 and 6 illustrate the corrosion pitting which had occurred; Figs. 7 and 8 typify the corrosion and poor penetration of the welds.

Intergranular sulphur penetration was sought specifically at high power but none was found. However, all specimens showed carbide precipitation and heavy etching attack at the grain boundaries (Fig. 9). This condition indicated that the material had been sensitized and was susceptible to aqueous intergranular corrosion.

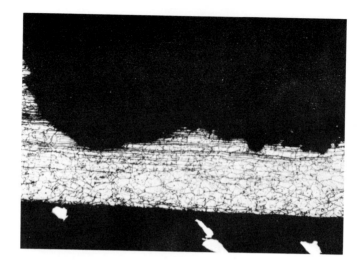

Fig. 5

Fig. 6

Figs. 5 and 6. Examples of corrosion pitting in plate. Etched electrolytically 5% HSO₄. 20×

Fig. 7

Fig. 8

Figs. 7 and 8. Examples of corrosion and poor penetration respectively of the welds. 20 ×

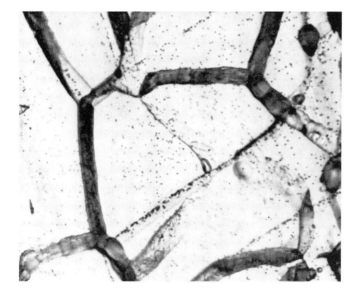

Fig. 9. Showing sensitised structure of plate. 750 ×

Microprobe analysis showed that the corrosion product contained sulphur, chlorine and calcium. The majority of the sulphur will have originated in the flue gas and the chlorine and calcium in the sea-water. Figure 10 is a stereoscan photograph of the corroded surface showing white spherical particles (S, Cr, Fe and Ni rich) and white angular particles (S, Ca and Fe rich).

The predominant acid radical in both the corrosion product and the carbonaceous deposit was sulphate with the concentration in the latter 10 × that of the former. This suggests that the corrosion was caused by sulphuric acid formed by oxidation of the SO_2 in the flue gas. The carbonaceous deposit will absorb some of the flue gas and release sulphurous and sulphuric acids onto the metal surface when moisture condensation occurs. The attack will be further accentuated by the carbon acting as a cathode in a galvanic cell with the metal surface and the susceptibility of the alloy to intergranular corrosion. The latter will facilitate the initiation and growth of corrosion pits at the grain boundaries.

Fig. 10. Corroded inside surface showing spherical and angular particles. 470 ×

Summary

Perforation of the inlet foot was caused by sulphuric acid formed by oxidation of the SO_2 in the flue gas and accentuated by the carbon deposit on the internal surfaces of the foot. Pitting corrosion was encouraged by the material being sensitized and susceptible to intergranular corrosion.

Corroded Pump Impeller

Friedrich Karl Naumann and Ferdinand Spies

Max-Planck-Institut für Eisenforschung
Düsseldorf

A cast iron pump impeller showed strong corrosion after an operating period of only one-half year. As can be seen from the section in fig. 1 the corrosion had penetrated substantially the wall thickness of the thin cross sections without loss of material. The pump had moved scrubbing water from a gas generator. According to an analysis by the City of Düsseldorf Waterworks the water had the following properties:

p_H-Wert/p_H value	7,2	
m-Wert/m value	3,13	
Gesamthärte/total hardness	32,4	°dH
Karbonathärte/carbonate hardness	8,8	°dH
Nichtkarbonathärte/non-carbonate hardness	23,6	°dH
Gesamtkohlensäure/total carbonic acid	74,3	mg/l
Gebundene Kohlensäure/combined carbonic acid	68,8	mg/l
Freie Kohlensäure/free carbonic acid	5,5	mg/l
Chlorid/chloride	99,4	mg/l
Sulfat/sulfate	378,5	mg/l
Eisen/iron	2,7	mg/l
Sulfid/sulfide	Spuren/traces	

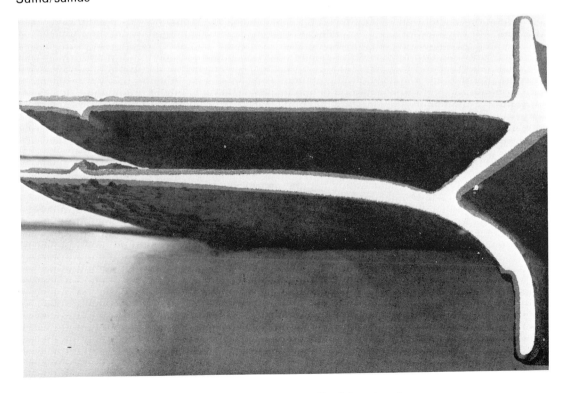

Fig. 1. Section through pump impeller. Graphitic corrosion designated by darker edges. 1 ×

Fig. 2. Thick-walled section

The impeller as well as the pump housing consisted of cast iron alloyed with nickel. Chemical analysis of the non-corroded part of the impeller showed the following composition [in wt. %]:

$C_{ges./total}$	Graphit/graphite
3,14	2,55
Ni	S
3,12	0,15

The microstructure consisted of lamellar graphite in a purely pearlitic matrix (fig. 2) in the thicker cross sections. In the thinner cross sections, the graphite had precipitated to a large extent in a fine-leafed structure by supercooled solidification (granular graphite), and the matrix was predominantly ferritic in these regions (fig. 3). This supercooled structure imparts only low strength to the iron, but should not affect the corrosion resistance.

In the attacked zone, corrosion took place along the phase boundary graphite/matrix (fig. 4) in the normal structure as well as in the supercooled one. It oxidized the ferrite

Fig. 3. Thin-walled section

Figs. 2 and 3. Core structure: Etch: Picral. 500 ×

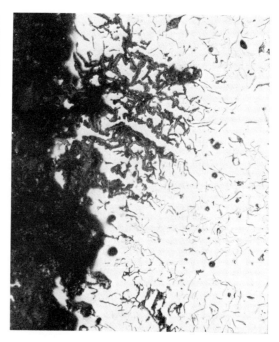

Fig. 4. Structure in area between attacked layer and unattacked core material. Unetched section. 100 ×

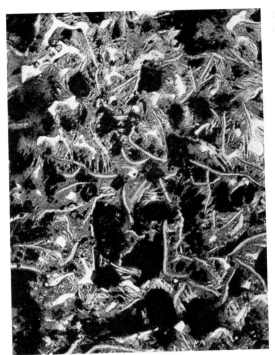

Fig. 5. Structure of the attacked layer. Unetched section. 200×

and pearlite of the matrix selectively while the graphite, the ternary phosphide eutectic steadide, and the sulfidic inclusions were not affected (fig. 5). This selective form of cast iron corrosion which is known by the de-

signation "spongiosis" or "graphitic corrosion" does not cause material loss, but lowers the strength of the cast iron to such an extent that a knife can cut or scrape off a black powder[1]. The powder thus produced contained in this case 10.85% C, 1.8% S and 1.45% P.

It is known that graphitic corrosion can occur through the effect of salt solutions or weak acids such as, for instance, from acidic soil or hydrogen sulfide-containing water. It appears possible that in the scrubbing water transported from the generator at least a part of the sulfate found was actually sulfide or hydrogen sulfide and was primarily responsible for the corrosion. Normally graphitic corrosion is a slow process[2] [3]. Comparable occurrences were found, however, in a relatively short time during tests to isolate electrolytically graphite, phosphide, carbides and non-metallic inclusions in the cast iron. Therefore it is reasonable to assume that in the failure at hand corrosion was accelerated by galvanic currents.

References

[1] F. K. NAUMANN, F. SPIES, Prakt. Metallographie 4 (1967) 367/370

[2] O. BAUER, O. KRÖHNKE, G. MASING, Die Korrosion metallischer Werkstoffe, Bd. 1 (1936) 280/283

[3] F. TÖDT, Korrosion u. Korrosionsschutz (1955) 125

Corroded Pipes from Gas Generating Plant

Friedrich Karl Naumann and Ferdinand Spies
Max-Planck-Institut für Eisenforschung
Düsseldorf

Pipes from a gas generating plant producing a gas containing 6% CO_2, 20% CO, 8 to 12% H_2, 0.5 to 1.5% CH_4, remainder N_2, were so heavily eroded after only four months that they had to be replaced. The temperature was 400 to 500°C. The pipes were supposed to have been made from a chromium-molybdenum steel, probably a high temperature steel with low chromium content. The object of the investigation was to determine whether the damage was due to mechanical wear or corrosion.

Three sections of pipe were examined. Specimen 1 from the inlet pipe of the hot gas valve had a light coloured deposit on the external surface which was probably fur while its inside surface was lined with a thin layer of black corrosion products, probably iron sulphide. Specimen 2, from the outlet pipe of the dust extractor had been reduced to a thickness of several tenths of a millimetre, apparently preferentially from the inside. Residues of a black corrosion product remained adhering to this specimen also. Specimen 3, taken from the pipe in the dust extractor was completely corroded apart from a few places along the welded seam. The black corrosion product had a lamellar appearance.

A sample of the corrosion product was scratched off all the specimens and analysed for carbon, sulphur and iron. The following values were obtained:

Specimen	C %	S %	Fe %
1	9,1	25,2	48,5
2	5,7	26,5	50,9
3	5,1	25,1	45,1

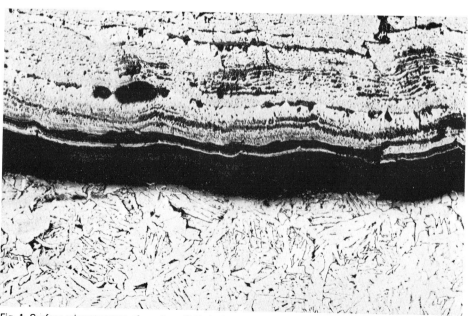

Fig. 1. Surface microstructure of specimen 2. Etched with alcoholic picric acid. 100 × (top: corrosion product, bottom steel)

According to these, the corrosion product consisted mainly of iron sulphide mixed with soot and rust.

The pipes had thus been eroded not by mechanical wear but by corrosion due to the high content of hydrogen sulphide in the gas. This was confirmed by metallographic examination. Figure 1 shows a section of the iron sulphide layer on specimen 2. The microstructure of the pipe was ferritic-pearlitic and the carbide of the pearlite was spheroidized. No conclusions could be drawn about the operating temperature because no information was available about the initial microstructure of the pipe. It is known, however, that hydrogen sulphide attacks iron strongly at 400 to 500°C with the formation of sulphide.

It can therefore be concluded that for the processing of the high sulphur content coal in question, the pipes must be made of a heat resistant steel with a higher chromium content and more resistant to attack by hydrogen sulphide than the material used here[1].

References

1) F. K. NAUMANN, Stähle für Treibstoffgewinnungsanlagen. Chem. Fabrik 11 (1938) Nr. 31/32, 365/376

Fractured Marine Riser

Fulmer Research Institute Ltd.

Buckinghamshire, England

Failure occurred in the connector groove of a marine riser coupling from a drilling rig. The steel specified for this component was AISI 4142 (0.40 to 0.45 % C; 0.75 to 1.00 % Mn; 0.20 to 0.35 % Si; 0.80 to 1.10 % Cr; 0.15 to 0.25 % Mo) normalised from 900°C.

The general appearance of the failed coupling is shown in Fig. 1. The fracture surface was heavily corroded due to its exposure to sea water but nevertheless the directions of crack propagation could be determined from the orientation of the clearly defined chevron markings. These chevron markings are characteristic of a fast-running brittle crack and their appearance at position C in Fig. 1 is shown in detail in Fig. 2. This indicates that at this position the crack was propagating in an anti-clockwise direction. Similar fractographic features were evident at position D, but in this case the chevrons indicated a clockwise direction of crack propagation. This indicates that the fracture initiated in the region of position A, then propagated as two cracks running simultaneously in clockwise and anticlockwise directions, eventually meeting opposite the point of initiation forming a pronounced step on the fracture surface at point B in Fig. 1.

Examination of region A revealed that the crack had initiated on the outside surface of the riser at the bottom corner of a circumferential connector groove. Having passed through the 12.7 mm ($^1/_2$ in) reduced wall thickness at the base of the connector groove, the two cracks then propagated into the full 25.4 mm (1 in) thickness, but returned to the connector groove at positions C and D. Propagation through the thicker section of the plate is not unusual since, due to the increased stress triaxiality, this can be a lower energy fracture path.

Detailed examination of the initiation region at position A revealed a discoloured area on the fracture surface which extended for approximately 65 mm (2.5 in) along the surface of the connector groove and penetrated to a maximum depth of 5.5 mm (0.25 in) into the fracture surface. This area is delineated in Fig. 3 and appeared to consist of a compact oxide scale. The same area is shown in Fig. 4 after chemical stripping of the corrosion

Fig. 1. General appearance of failed coupling

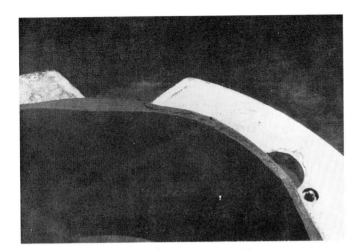

Fig. 2. Showing chevron markings on fracture surface at position C (Fig. 1)

products, and at least two separate initiation sites can be distinguished. Also evident in Fig. 4 are numerous corrosion pits, 0.25 mm to 0.125 mm deep — on the inclined face of the groove.

Sections of the fracture surface were examined in the scanning electron microscope after chemical stripping of the corrosion products with a solution 50/50 HCl/water inhibited with hexamine. The resultant clean-

Fig. 3. Fracture surface at position A (Fig. 1), illustrating compact oxide scale region within dotted line

Fig. 4. Same area as Fig. 3, after chemical stripping, showing (arrowed) two separate initiation sites

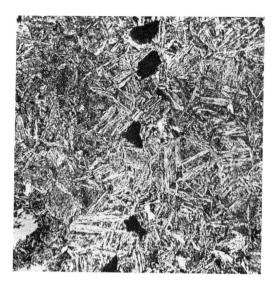

Fig. 5. Upper bainitic structure with aligned grains of unresolved pearlite (black). 100 ×

Fig. 6. Brittle cleavage surface of fracture toughness specimen. 500 ×

ed surfaces revealed that the region of fast fracture consisted entirely of brittle cleavage failure. The initiation area was more heavily corroded but some cleavage facets could be distinguished in addition to a generally rather featureless fracture appearance. This suggests that both corrosion fatigue and stress corrosion may have contributed to the initial stage of slow crack growth.

Sections for optical micro-examination taken from areas both adjacent to and remote from the fracture had similar micro-structures consisting of upper bainite together with grains of unresolvable pearlite aligned in bands along the longitudinal direction of the pipe. The pearlite grains occur in lean alloy microsegregation bands where local hardenability is low. The structure, shown in Fig. 5, is typical of a low alloy hardenable steel in the normalised condition.

This type of microstructure is very brittle and although tensile tests confirmed that the steel satisfied the specified strength requirements (700 MNm^{-2} Y. S.), impact tests revealed a fracture transition temperature in excess of + 100°C. Figure 6 shows the completely brittle cleavage mode of fracture exhibited by a fracture toughness specimen.

Summary

From the results of visual examination, optical and scanning electron microscopy, it has been possible to draw up a sequence of events leading to failure:

i) Formation of corrosion pits in the connector groove due to the presence of sea water.

ii) Initiation of a corrosion fatigue crack in the connector groove.

iii) Fatigue crack propagation in a direction normal to the axis of the coupling, probably subsequently assisted by stress corrosion.

iv) After reaching a crack depth of 5.5 mm, rapid unstable brittle fracture occurred resulting in catastrophic failure in the marine riser.

The main recommendation was to specify a quench and temper, rather than a normalising treatment which would produce a high toughness steel without loss in strength. This recommendation would be suitable for existing couplings but the use of a lower carbon quenched and tempered steel for new couplings would not only give a very high toughness, resistance to stress corrosion cracking and pitting corrosion, but would also be very much more easily welded than the present low alloy steel.

Corroded Leaky Stainless Steel Pipes

Friedrich Karl Naumann and Ferdinand Spies

Max-Planck-Institut für Eisenforschung

Düsseldorf

Pipes of 23 mm diameter and 2 mm wall thickness made of 18/8 steel started to leak through corrosion after 1½ years' use in a soft water preheater of a food factory. The pipes served to heat from outside to 100°C completely desalinated, virtually oxygen-saturated water by steam of 0.5 atm. and 150°C.

The pipes showed on the outside a reddish-brown coating with a few flat pitting holes and incipient cracks (Fig. 1). The cracks were markedly widened in 180° bends (Fig. 2). The interior of the pipes was covered with rust patches (Fig. 3) under which deep and partially penetrating pitting scars were visible (Fig. 4). From these scars numerous cracks branched out that were predominantly transcrystalline (Fig. 5). Therefore the leaking of the pipes is caused by pitting and stress corrosion. These types of corrosion often occur together.

Chlorides are the most likely corrosive agents for this type of attack, while concentrated alkaline solutions are somewhat less corrosive. If no chloride ions were present in small concentration in the water treated in the pipes, this case may show that other possibilities exist as well. Stresses were present in the form of residual stresses from drawing and bending of the pipes.

Fig. 1. Outer face of pipe with flat pitting holes and incipient cracks

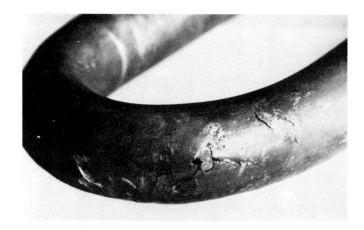

Fig. 2. 180° bend with markedly widened cracks

It was recommended to the factory to use in the future molybdenum alloyed steels of the type 18 10 or 18 12. These ar more resistant against local disruption of the passive film and pitting than the molybdenum-free 18/8 steel used here.

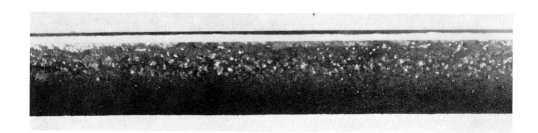

Fig. 3. Rust patches on interior side of the pipes

Fig. 4. Cross section through pipe with pitting scars

Fig. 5. Transcrystalline cracks originating from pitting scars, cross section, unetched

Leaking Coil Made of Stainless Steel

Friedrich Karl Naumann and Ferdinand Spies
Max-Planck-Institut für Eisenforschung
Düsseldorf

A cooling coil made of austenitic stainless steel (chromium-nickel-molybdenum X 10 CrNiMoTi 18 10, material No. 1.4571) started leaking in 15 spots after 8 weeks' service in an apparatus in which ammonium sulfide solution containing approx. 7.5% H_2S and 14% NH_3 was converted into ammonium sulfate solution under pressure of 25 to 30 atm. with addition of air. This coil was cooled by water at 3 atm. and its external temperature was approx. 175°C. All other parts of the apparatus that were made of the same steel remained tight and clean.

Two sections of the coil were examined. They showed pinhead size pitting cavities at the exterior surface and fine cracks, appearing in large arrays that were partially parallel and partially at an angle to the longitudinal direction of the pipes and which appeared at the external as well as the internal surfaces of the bend. The pipe diameter was flattened in the bends.

Chemical analysis showed the following values in wt.-%:

C	Cr	Ni
0,17	17,35	11,45

Mo	Ti
2,00	0,44

Accordingly carbon content was too high. It will be shown later that this is a consequence of the carburization of an internal surface. The other values corresponded to the standard.

Metallographic examination was conducted on longitudinal and transverse sections. They showed that the cracks originated at the external surfaces of the pipe. Individual cracks had already penetrated the entire wall (Fig. 1). They had branched out considerably

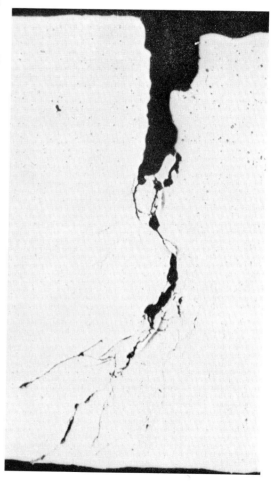

Fig. 1. Traversing crack (exterior of pipe, top). Unetched transverse section. 45 ×

and were predominantly transcrystalline (Fig. 2). Many originated from pits (Fig. 3). According to their appearance they are stress corrosion cracks, that occur in these austenitic steels under the combined effect of tensile stresses, be these external stresses or residual stresses, and certain corrosion agents, especially chlorides. The interior surfaces of the pipes were not attacked.

The dark peripheral zone proved to be carburization (Fig. 4) when observed under high magnification. This may have been caused by the use of a carbon-containing filler during hot bending of the pipes. It was of no consequence to the damage.

The type of corrosion agent and the stresses can only be surmised. If chlorides were absent, which could not be established, hydrogen sulfide may be suspected which similarly causes pitting and in high strength steels is also capable of causing cracks[1]).

Stresses may be residual or due to such operations as heat treating, bending or straightening. The better behavior of the container lining made of the same steel may be explained by a more favorable state of stress of the sheets used.

References

1) F. K. NAUMANN, W. CARIUS, Bruchbildung an Stählen bei Einwirkung von Schwefelwasserstoffwasser, Archiv Eisenhüttenwes. 20 (1959) 233/238

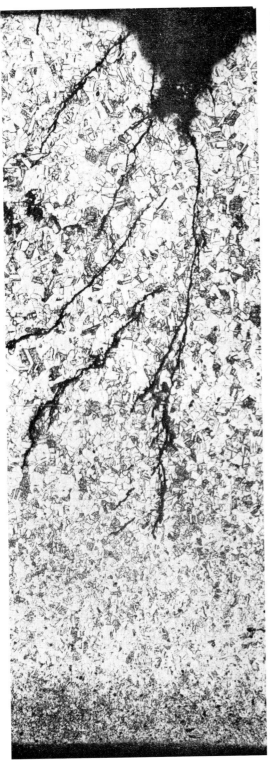

Fig. 2. Stress corrosion cracks originating at exterior surface. Carburized zone at internal surface, transverse section, etch: V2A-Pickle. 90 ×

(Continued on the next page)

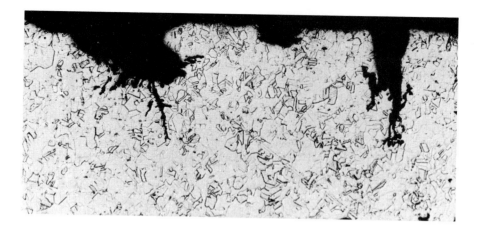

Fig. 3. Pitting corrosion with incipient crack formation. Longitudinal section, etch V2A Pickle. 100 ×

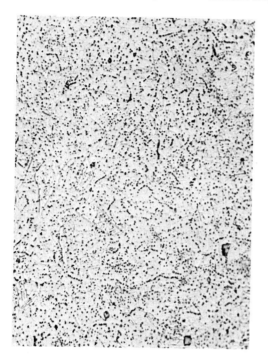

Fig. 4. Structure of carburized zone at the internal surface. Longitudinal section, etch: V2A-pickle. 500 ×

Fractured Lock Ring of Drum

Friedrich Karl Naumann and Ferdinand Spies

Max-Planck-Institut für Eisenforschung
Düsseldorf

The lock ring of a centrifuge drum that turned at 4500 rpm fractured after one year's operation. The ring had a trapezoidal thread on the inside whose uppermost turn can be seen in Fig. 1 (G). It had fractured radially in one of four places in which the cross section was weakened by short grooves (N) that apparently had served as tool grips for tightening the cover. The ring was made of steel with approximately 0.5 % C, 1.3 % Mn and 1.1 % Cr and was hardened and tempered to 105 kp/mm² strength at 11 % elongation (δ_{10}). Notch impact strength was 1.5 to 2.5 kpm/cm² (DVM), a comparatively low value which may be caused by the sensitivity of this steel to embrittlement during slow cooling after tempering.

The fracture apparently propagated from the base of the thread and then followed the thread in a circumferential direction (Figs. 1 and 2 horizontal at bottom). It then broke through radially at the top across the ring due to a weakening caused by the external reduction of the cross section. The uppermost turn that was still present in the fractured specimen sent in for examination was corroded at the base by pitting (Figs. 1 b and 2). Corrosion was favored in this case by differences in ventilation and formation of so-called Evans elements in the narrow gap between thread and counterthread.

Metallographic examination showed that such pitting favored intergranular fissures of minor depth (Figs. 3 and 4). Therefore stress corrosion seems to be clearly established. It evidently accelerated cracking of the ring.

Since the drum was used for the processing of various liquids, we cannot state which medium caused the corrosion.

Fig. 1 a. Transverse fracture

Fig. 1 b. Inner surface at transverse fracture

Figs. 1 a and b. Views. N = groove at exterior, G = uppermost thread inside. Fracture descaled with dilute hydrochloric acid with inhibitor. Approx. 1 x

Fig. 2. As Fig. 1 b. Bottom: circumferential fracture. 3 x

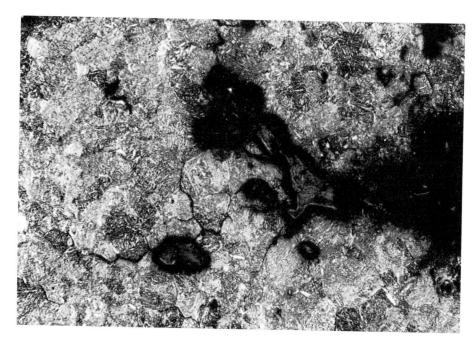

Fig. 3. Corrosion pit with intergranular cracks

Fig. 4. Stress corrosion crack propagation

Figs. 3 and 4. Section, etch: Nital. 100 x

Intercrystalline Corrosion of Welded Stainless Steel Pipelines in Marine Environment

Emanuele Mor, Eugenio Traverso and Giovanna Ventura

Laboratorio per la Corrosione Marina dei Metalli, C.N.R.

Genova, Italy

Introduction

By studying the corrosion on industrial systems, it often happens to meet process and etching figures which could be surely avoided by a more accurate evaluation of the aggressive environment and a more precise material selection. Besides, an installation incompetence is sometimes added to these basic defects.

The present article is just a practical case of this type, where some defective weldings are added to the improper choice of a stainless steel for water pipeline.

Comparing the results obtained by different investigations reported in the literature it can be concluded that the causes for the steel susceptibility to intercrystalline corrosion may be quite different, but all of them involve precipitation at the grain boundaries of a surplus phase (non-equilibrium or equilibrium carbides or other phases from γ solid solution).

In particular, the austenitic stainless steels may show a considerable susceptibility to intercrystalline corrosion in such conditions as, for example, during welding.

In fact, if these steels are heated within a temperature range from 425 °C to 870 °C (known as sensitization range) and slowly cooled, the carbon, in γ phase solid solution, precipitates preferentially at the grain boundaries where it combines with the chromium to make carbides rich in this element, probably of the type $M_{23}C_6$ [1] [2]).

In a normal welding process, necessarily, a thermal gradient between the weld metal and the parent material sets up, which decreases to zero with the increasing distance from the joint. In the area subjected to this gradient will exist, therefore, a narrow band of parent metal or seam formerly deposited, which was heated in the temperature range from 425 °C to 870 °C for a time long enough to promote the intergranular carbide precipitation [3] to [6]).

At these temperatures, in fact, the carbon may diffuse rapidly towards the grain boundaries, while the chromium, less mobile, migrates very slowly. Therefore, the carbon combines with the chromium, which is immediately adjacent to the grain boundaries, and produces a region whose chromium content may also decrease below 12 wt.-% (the minimum chromium content so that a steel may be considered a stainless steel).

According to this, a potential difference sets up between the grains, constituting large cathodic areas, and the chromium depleted areas, immediately adjacent to the grain boundaries, constituting little anodic areas.

Such a fact leads to a rapid grain boundary attack and to a comparatively weak corrosion of the same grains [7] [9]) and that just results in a typical intercrystalline corrosion. The carbide precipitation rate is related to the steel carbon content.

In the 18/8 stainless steels, the carbon is stable in solid solution, on all conditions, up to a maximum content of 0,03 wt.-%. With the increasing of this content, up to about 0,08 wt.-%, the precipitated carbide rate increases very slowly and is often insufficient to promote a real reduction of the corrosion resistance.

With higher carbon contents, on the contrary, the carbide precipitation increases very fast [4] [5] [7] [9] [13]). So, the low-carbon stainless steels are preferred in the welded systems, subjected to strong corrosive media causing this type of corrosion as, for example, sea water, systems which cannot, for practical reasons, be subjected to heat-treatments.

To prevent the occurrence of this phenomenon, the literature suggests a number of methods available for producing austenitic stainless steels, which can be stabilized against intercrystalline corrosion [7] [10] to [12].

The more common are:

1. A reduction of the carbon content of the steel

The aim of this technique is to reduce the amount of $M_{23}C_6$ at the austenitic grain boundaries. This method is the basis of some commercial alloys as, for example, the extra-low carbon stainless steels (types AISI 304L and AISI 316L).

2. An appropriate heat-treatment at a temperature high enough to dissolve the carbides (usually from 1037 °C to 1150 °C) and for a regulated time (to avoid an excessive grain growth), followed by quenching

Localized heat — treatment of the area immediately adjacent to the weld, is not, in fact, satisfactory to prevent the chromium carbide precipitation; for an effective heat-treatment the entire unit must be heated and quenched.

A heat-treatment followed by quenching from 800 °C to 900 °C would be also possible to isolate the carbides and to avoid a continuous path of the sensitized zone at the grain boundaries. It should prevent a continuous penetration path for the corrosive medium [8]. Such a treatment has a limited commercial application.

3. A cold work of the steel before the sensitization

The cold work promotes, in fact, the chromium carbide precipitation on dislocations by hardening and, therefore, reduces the carbide amount at the grain boundaries. However, this treatment is necessarily restricted to a certain form of material such as strip, sheet or wire and also precludes the use of subsequent treatments at temperatures higher than that of sensitizing treatment.

4. A reduction of the austenitic grain size

It has been found that coarse-grained steels are more susceptible to intergranular corrosion than fine-grained steels, with the same composition [14]. However, difficulties concerning the regulating of the grain size, particularly in welded materials, make this technique also limited in commercial use.

5. An addition of stabilizing elements to precipitate the carbon as a stable carbide

An element will be effective as a stabilizer if it has a higher affinity to the carbon than the chromium and if, at the temperatures in the sensitizing range, it is able to form a carbide less soluble in the γ phase than the chromium carbide.

The most used stabilizing elements are titanium and niobium (sometimes also Nb plus Ta).

The titanium additions should be, in theory, five times the carbon content of the steel, but, in practice, owing to the titanium tendency to combine with other elements, particularly nitrogen and oxygen during the welding, the Ti:C ratio must be greatly higher (sometimes up to 7,5:1) [7].

Similar considerations can be made for the niobium whose amount must be not smaller than eight to ten times the carbon content [16].

Examples of steels satisfying these conditions are the types AISI 321 and 347, the alloy 20 (high-nickel stainless steel) and the alloy Ni-Fe-Cr 825 [4] [17].

The austenitic stainless steels are reported to derive their corrosion resistance from a passive oxide film, which forms on their surface.

In oxidizing media (such as the sea water with 5 to 10 mg/l oxygen dissolved) the film remains intact, but can be broken in the regions where the oxygen concentration is insufficient.

The oxygen availability in some localized areas can be slackened or prevented by interstices or deposits formed on the metallic surface. The uncovered metal will act as a soluble macro-anode compared to the surrounding large zone covered by the oxide passive film (cathodic area) in the same way as it happens, at micro-level, between the grain boundaries and the grain.

The macro-galvanic actions, due to the oxygen concentration cells, will promote, inevitably, some localized pits, where the attack rate will result strongly favoured by the high ratio between the cathodic and anodic areas.

In these cases, the potential difference can attain a value of 500 mV and exceed the values reported in literature references [19]).

The stainless steels, at the passive state, act, in fact, in sea water as a noble metal. As can be seen in Table 1 the AISI 304 and 316 potentials, in the passive state, are respectively 0,08 V and 0,05 V negative with respect to saturated calomel electrode.

Under the experimental conditions given in Table 1, the potential differences between the active state (in our case the breakdown of the protective oxide film occurs because of the oxygen absence) and the passive state result in 0,45 V and 0,13 V, respectively, for AISI 304 and AISI 316.

Table 1. Galvanic series in moving sea water (4 m/sec) at 25 °C [19])

Element or alloy	Potential (vs. C. E.)
Zinc	− 1,03
Aluminium 3003	− 0,94
Carbon steel	− 0,61
Cast iron	− 0,61
AISI 304 (active)	− 0,53
Copper	− 0,36
Admiralty	− 0,29
Cupro-nickel 70/30 (0,47 Fe)	− 0,25
Nickel 200	− 0,20
AISI 316 (active)	− 0,18
Inconel 600	− 0,17
Titanium	− 0,15
Silver	− 0,13
AISI 304 (passive)	− 0,08
Monel 400	− 0,08
AISI 316 (passive)	− 0,05

In stagnant sea water, however, in presence of deep interstices, as in our welded pipeline, these potentials can attain values considerably higher than above reported, especially in absence of molybdenum.

As established in practice, the molybdenum, as alloying element, has a significant beneficial effect on overall corrosion behaviour by lowering the pitting frequence even if the pitting depth seems not to be affected in the same way. The advantage of the molybdenum presence should be associated with its effect on the probability of the localized attack, rather than on its intensity.

No doubt that, as reported from literature [3]), the sea water movement, at least from 1,20 to 1,50 m/sec, would have produced conditions less favourable to the setting up of the above mentioned galvanic couples. The defects, which we have found, are, therefore, easily attributed to the permanence of the stagnant water.

The iron sexquioxide layers, deposited on the internal pipeline surface, are an evidence of it (Fig. 2).

Experimental

The pipe specimen shown in Figs. 1 and 2 was cut from a pipeline system for the feed of a swimming pool with sea water. The pipeline was in intermittent service for about two years when a failure in a transversal weld took place.

Metallographic examinations of the specimen showed the presence of macroscopic working defects in the transversal welding seam, such as lack of root penetration, inadequate preparation of the edges to be welded and the presence of electrode slags (Fig. 3). It is clear that the latter prove that the welding was made with the coated electrode process, which, from experience, is known to be the least suitable for a good penetration into the transversal joints of stainless steel pipes.

The numerous points of corrosion initiation visible in the sensitized area adjacent to the longitudinal weld, are of great interest with respect to the transversal weld, where on the other hand the defects are connected to the lacking manual dexterity (Fig. 1).

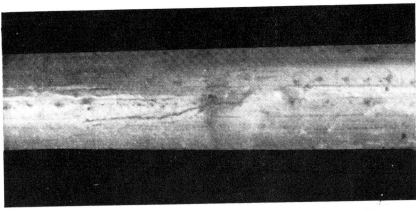

Fig. 1. Longitudinal welding seam (external side). Start points of corrosion. 0.4 x

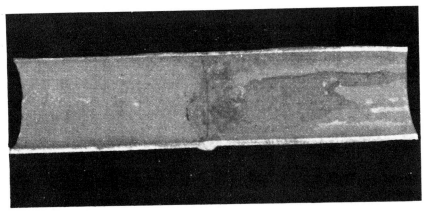

Fig. 2. Transverse welding seam (internal side). 0.4 x

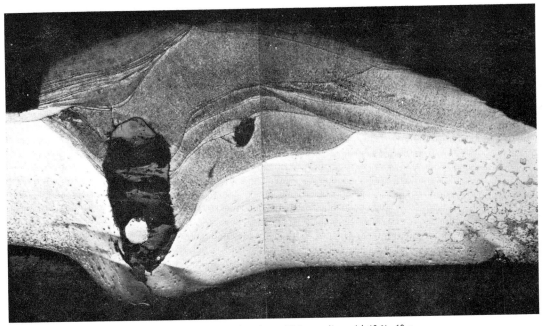

Fig. 3. Section of the transverse weld. Electrode slags. Etch: oxalic acid 10 %. 10 x

Fig. 4. Cavity in the sensitized area. 10 x

A transverse section of this area showed a deep cavity at about 3 mm from the weld metal, corresponding to the external points of attack (Fig. 4).

The microscopic examination of this hole showed the grain boundaries to be separated (Fig. 5), allowing to decide the intergranular type of the corrosion, due to a precipitation of carbides. These carbides are shown in Fig. 7 at higher magnification after chemical etching.

Figure 6 also corroborates the preferential attack at the grain boundaries and shows that some grains are completely detached.

All that indicates, therefore, that the intermediate region, between the weld metal and the starting points of corrosion, was subjected during welding to higher temperatures than the critical range ($> 870 \,°C$). In fact, in this region no attack was present. The corroded zone, on the contrary, has been left in the thermal range $425 \,°C$ to $870 \,°C$ for a period sufficiently long to promote the precipitation of a considerable carbide amount.

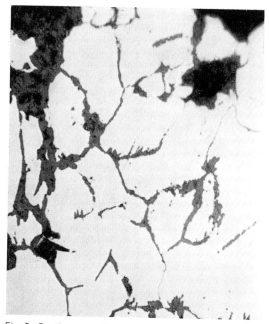

Fig. 5. Continuous path along the grain boundaries. 500 x

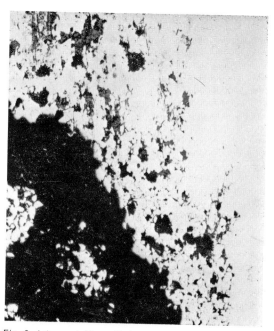

Fig. 6. Intercrystalline attack. Grains detached. 50 x

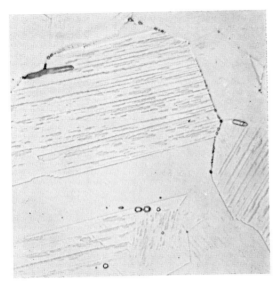

Fig. 7. Grain boundary carbides and creep lines in the sensitized area. 800 x

The electrolytic etching with oxalic acid pointed out the weld material (Fig. 8). The welding was made in two seams; the first penetrated completely the tube thickness, but without the melting of one of the edges, probably because of an insufficient preparation and misalignment.

The following seam (repair) was not able to penetrate into the root and to eliminate the unmelted edge section and the consequent corner.

The welded edge misalignment and the dissatisfactory penetration of the weld metal, also found in some other points of the seam (Fig. 9), produced a shape where the renewal of the electrolytic medium is insufficient and where the approach of oxygen to form a passivating film is also difficult. That produces anodic areas with lower potentials where the metal passes into solution, as above mentioned.

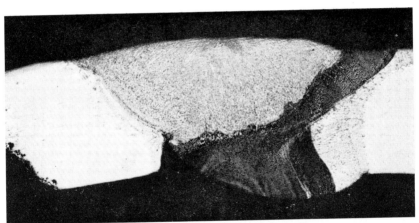

Fig. 8. Section of longitudinal weld. Etch: oxalic acid 10 %. 10 x

Fig. 9. Section of longitudinal weld. Edge misalignment. Etch: oxalic acid 10 %. 10 x

Fig. 10. Detail from Fig. 9. 80 x

The second seam, moreover, sensibilized the previous one and increased the susceptibility to a corrosive attack both on the parent metal and on the weld material (Fig. 10).

In the latter, the situation is made worse by the presence of micro-cracks due to very high localized heat concentrations with consequent thermal expansion and considerable shrinkages during the cooling.

Chemical analysis permitted to classify the material in the not-stabilized austenitic stainless steel series ($> 0,08$ wt.-% C).

In Table 2, the chemical analysis of the above mentioned material is compared to the most frequently used stainless steels in marine environment.

By comparing these compositions, the unsuitable choice of the steel for this system, is evident above all owing to the carbon content being higher than standard values (0,03 wt.-%) and not being compensated by stabilizing elements.

Summary

The authors examined a stainless steel pipeline, used in a marine environment, which showed decay along the weldings. These defects were evidenced by means of metallographic and electrochemical examination.

The probable causes of the corrosion and the precautions to prevent these processes are examined.

It is possible, therefore, to conclude that some of the corrosion processes may be avoided by a more accurate selection of materials with respect to the aggressive medium to be used.

Table 2. Chemical composition of stainless steels used in marine environments

Content [wt %]	C	Mn max	P max	S max	Si max	Cr	Ni	Mo	Other
Specimen/Probe	0.103	1,24	0,016	0,014	0,65	17,23	7,21	0,22	0.12 Cu
AISI 304L	0,03max	2,00	0,045	0,030	1,00	18–20	8–12	–	–
AISI 316L	0,03max	2,00	0,045	0,030	1,00	16–18	10–14	2–3	–
AISI 321	0,08max	2,00	0,045	0,030	1,00	17–19	9–12	–	Ti5xCmin

References

1) T. G. GOOCH, Br. Weld. J. 15 (1968) 345
2) T. G. GOOCH, Metal Constr. 1 (1969) 569
3) H. H. UHLIG, The Corrosion Handbook, 4th edn, J. Wiley and Sons, New York (1953) 161
4) F. N. SPELLER, Corrosion Causes and Prevention, 3rd edn, McGraw-Hill Book Company, New York-London (1951 227
5) T. H. ROGERS, Marine Corrosion, G. Newnes, London (1968) 81
6) E. KAUCZOR, Der Praktiker 1 (1972) 9
7) F. G. WILSON, Br. Corros. J. 6 (1971) 100
8) A. JOSHI, D. F. STEIN, Corrosion NACE 28 (1972) 321
9) V. CIHAL, Corrosion Traitments and Protection 18 (1970) 441
10) Metals Handbook, 8th edn, Am. Soc. for Metals, Novelty (1961) 565
11) T. H. ROGERS, The Marine Corrosion Handbook, McGraw-Hill Company of Canada Limited Toronto (1960) 262
12) R. BOULISSET, J. DOLLET, Metaux Corrosion Industrie 579 (1973) 375
13) I. A. LEVIN, Proceedings of the third International Congress on Metallic Corrosion, vol. II, Moscow (1966) 418
14) C. P. DOSHI, W. W. AUSTIN, Corrosion 21 (1965) 332
15) T. G. GOOCH, J. HONEYCOMBE, P. WALKER, Br. Corros. J. 6 (1971) 148
16) M. S. GONCHAREVSKY, L. P. SCHESNO, Proceedings of the thirth International Congress on Metallic Corrosion, vol. II, Moscow (1966) 428
17) A. H. TUTHILL, C. M. SCHILL, C. M. SCHILL-MOLLER, The Ocean Science and Ocean Engineering Conference, Washington (1965)
18) A. P. BOND, H. J. DUNDAS, W. SCHMIDT, M. WOLF, Revue de Métallurgie (1973) 41
19) F. L. LAQUE, Corr. Testing Proc. ASTM 51 (1951) 495

Examination of Steel Specimens from an Ammonia Synthesis Installation

Friedrich Karl Naumann and Ferdinand Spies
Max-Planck-Institut für Eisenforschung
Düsseldorf

Unalloyed steels and the pure nickel steels frequently used in the past for highly stressed forgings are attacked by hydrogen under high pressure. The attack causes decarburization that leads to a loosening of the structure through precipitation of methane on the grain boundaries[1]. The critical temperature at which the attack occurs lies between 200 and 300°C, depending upon the hydrogen pressure[2]. Parts of an apparatus that are stressed in this temperature region must be checked constantly if they are not made from hydrogen-resistant steel[3]. Comparatively minor occasional temperature peaks may lead to attacks of dire consequences. In the following two examples taken from such control tests are described, especially because they illustrate the difference between hydrogen attack and oxygen decarburization.

1. A ring specimen was cut off from a high pressure line of unalloyed steel St 55.25 and was tested for hydrogen attack. The high pressure line was used for a gas mixture leaving the pressure vessel of an ammonia synthesis installation. The pressure of the installation was approx. 850 atm, while the temperature at this point was said to be 250°C. The gas consisted of approx. 69 vol.% hydrogen, 23 vol.% nitrogen and 8 vol.% ammonia. The pipe showed no hydrogen attack during macroscopic metallographic investigation. Tensile and notch impact tests showed normal mechanical values. Under the microscope, surface decarburization of several tenths millimeters depth could be detected

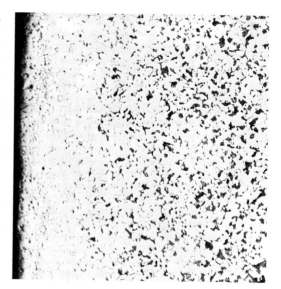

Fig. 1. Surface decarburization at the inner surface of a high pressure line, made of unalloyed steel. Transverse section, etch: Picral, 100×

(Fig. 1) under the inner surface and also, though less pronounced, under the outer surface. But the structure was completely dense. Therefore this was likely to be decarburization caused by oxidation during pipe rolling. This was confirmed by oxide precipitates in the decarburized zone (Fig. 2). This pipe was therefore not attacked. But the conditions to which these pipes were exposed are known to be close to the limit that an unalloyed steel can withstand, and therefore these lines were later replaced by chromium-molybdenum steel 20 CrMo 9 (Material No. 1.7283), a material that is reliably stable under these operating conditions. The unavoidable peripheral decarburization occurring during pipe manufacture however is present also in such pipes as shown in Fig. 3.

[1] F. K. NAUMANN, F. SPIES, Prakt. Metallographie 8 (1971) 375/384

[2] G. A. NELSON, Stahl und Eisen 80 (1960) 1134

[3] H. KIESSLER, Werkstoff-Handbuch Stahl u. Eisen, Düsseldorf (1965) Blatt 0 85

Fig. 2. Oxide precipitates in the decarburized zone of the pipe according to Fig. 1. Unetched transverse section. 200×

Fig. 3. Surface decarburization at the inner surface of a high pressure line of hydrogen resistant chromium-molybdenum steel 20 CrMo9. Transverse section etch: Nital 100×

2. During screwing of such a line into the closure head of the pressure vessel, which was made of nickel steel, an annular chip of the inner and seal plane of the closure piece was squeezed off. At this location similar conditions must have prevailed as in the adjoining pipe line with the exception of possibly somewhat higher temperatures. The chip was clearly attacked by hydrogen as was shown by metallographic examination. Figure 4 shows the structure of the inner surface of the closure piece. The grain boundaries are ripped open under the pressure of the precipitated methane, but decarburization that

leads to losening of the structure is not yet clearly visible because diffusion of the carbon at the grain boundaries at which the reaction takes place, occurs slowly at such low temperatures. Fig. 5 shows the normal structure of the closure piece at an unattacked spot for purposes of comparison. These high pressure vessels and closures were later also made of hydrogen-resistant steel.

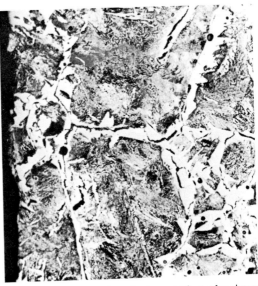

Fig. 4. Hydrogen attack at the inner surface of a closure piece of nickel steel. Section etched with Picral. 200×

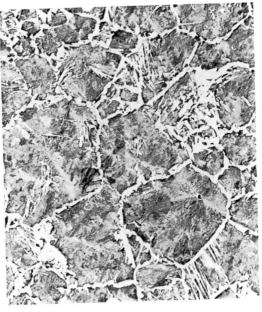

Fig. 5. Structue at the outer periphery of the closure piece according to Fig. 4. Section etched with Picral. 100×

Failure of Ball Joints

Fulmer Research Institute Ltd.
Buckinghamshire, England

For many years ball joints used in heavy vehicle steering joints had given satisfactory service. Suddenly, some failed a few hours after being fitted but before the vehicle had gone into service. Failure occurred in the necked part of the component between the thread and tapered section. The timing of these failures coincided with production problems which resulted in the joints being used immediately on finishing.

The ball joints are forged from En 353 (BS970 :815A16), machined, threadrolled, heat treated and finally ground. The heat treatment consists of carburising in a cyanide bath for 12 hours at 930°C, holding at 870°C for about 20 mins, then quenching in oil (carburisation of the threaded section is prevented by prior plating with copper). The ball joints are then tempered for 2 hrs at 170-175°C and finally the threaded end is immersed in acid for $\sim$ 45 mins. to remove the copper plate. When fitted, the ball joints are loaded to a torque of 200 lbs. ft. which produces an extension of 0.025 mm.

A longitudinal section through a failed ball joint is shown in Fig 1. The primary failure is on the extreme left of the micrograph and secondary cracking can be seen running from the machined surface normal to the major axis of the component. The light coloured area towards the top of the micrograph has been carburised and did not respond to etching. This effect, caused by lack of copper plating, is not thought to have contributed significantly to the failure.

The general structure of the pin was martensitic with a surface case depth of approximately 0.75 mm to 1.0 mm. The core and case hardnesses were 407 HV and 670 HV respectively. Analyses of the core material showed that the ball joint was within its specified chemical composition. Scanning electron microscopy of the fracture surfaces showed that the fractures were flat macroscopically; ridges radiating from the point of crack initiation (see Fig. 2) are characteristic of brittle

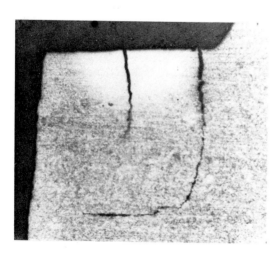

Fig. 1. Section through fractured ball showing secondary cracking. 25 ×

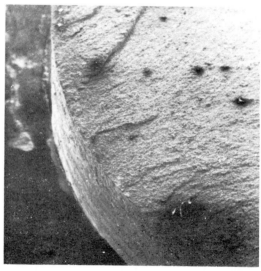

Fig. 2. Showing region of failure initiation. 10 ×

failure. Occasionally secondary cracks of approximately equal length can be distinguished running parallel to the primary fracture and these give rise to steps on the perimeter of the fracture surface opposite to the point of crack initiation (see Fig. 3). Examination at higher magnifications revealed that the fracture mode around the perimeter was completely intergranular (see Fig. 4) to a depth of about 2.5 mm which corresponds to the width of the stepped region in Fig. 3. Below this depth, the fracture mode changed abruptly to a finely dimpled fibrous fracture.

The most significant features of the failures are:

(i) Failure does not occur immediately but takes place some hours after loading.

(ii) The fracture is intergranular around the perimeter to a depth 2 to 3 times greater than the case depth; below this depth the fracture mode is fibrous.

(iii) Subsidiary cracks have been initiated at the surface of the component and all have propagated to approximately the same length.

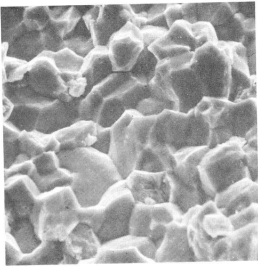

Fig. 4. Peripheral regions of fracture showing intergranular characteristics. 1300 ×

These features are characteristic of hydrogen embrittlement, the delayed nature of the cracking being due to the hydrogen slowly diffusing to regions of stress concentration where it promotes intergranular failure. The probable source of the hydrogen was the acid solution used for removing the blanking copper. Clearly, if this operation was carried out prior to the tempering, then the tempering operation would drive out the hydrogen thereby avoiding embrittlement problems no matter how quickly the ball joints are put into service after manufacture. The reason why previous samples hadn't failed was that during the normal course of production they were stored at room temperature for sufficient time to enable the hydrogen to have time to diffuse out before being fitted into the steering assembly.

Summary

The cause of failure was hydrogen, this being introduced into the steel during stripping of the copper and causing surface cracking when the component was stressed during fitting.

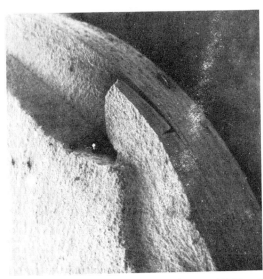

Fig. 3. Showing secondary cracks and stepped linking of final failure. 10 ×

Investigation of Distribution Manifolds from the Cooling Unit of an Ammonia Synthesis Plant

Friedrich Karl Naumann and Ferdinand Spies

Max-Planck-Institut für Eisenforschung

Düsseldorf

A cooler of an ammonia synthesis plant was destroyed after 3 years' service due to the rupture of a distribution manifold. The cooler consisted of a row of upper distribution manifolds and lower collection manifolds with a system of pipes between them. Synthesis gas under high pressure and at about 300° C, consisting of approx. 10 % NH_3 and unconverted gas of 25 % N_2 and 75 % H_2 content, was water-cooled externally to room temperature in this unit.

Figure 1 shows the fracture location and Fig. 2 the structure of the fracture. The fracture originated from inside the cross-boring for the gas inlet tube and has the typical flat-grey fibrous structure of a material destroyed by hydrogen [1]. Initial cracks in the crossboring in additional manifolds of this cooler, could also be observed or made visible by means of magnetic particle inspection (Fig. 3). The starting points of these fissures lay partly on the edge formed by the

Fig. 1. Sideview of the cracked cooler manifold. 0.4 x

Fig. 2. View of fracture of the cracked cooler manifold. 0.6 x

Fig. 3. Sideview of the cooler manifold with initial cracks. 0.6 x

Figs. 1 to 3. Views of the inlet side

Fig. 4. Interior view of the crossboring in Fig. 3. 1 x

Längsbohrung ⟶   ↑ *Querbohrung*

Fig. 5. Break of the opened crack in Fig. 4. 1 x

transverse and longitudinal boring and partly next to it in the longitudinal boring itself (Figs. 4 and 5). From this location the fissures entered the crossboring towards both sides.

Specimens for the metallographic investigation were removed parallel and at right angles to the fissures; these specimens cut into both the longitudinal and the transverse borings. Under both borings the structure appeared to have been loosened by intergranular separations (Figs. 6 to 8). These are the significant characteristics of hydrogen attack [1]). Decarburization had not yet become perceptible because of the small

Fig. 6. Initial cracks in the attacked zone. Etch: nital. 10 x

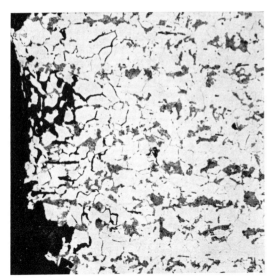

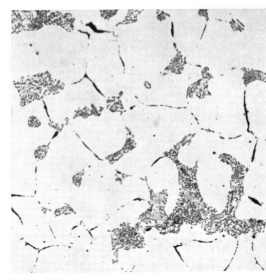

Fig. 7. 100 x

Fig. 8. 500 x

Figs. 7 and 8. Structure at the edge of the longitudinal boring in Fig. 5, showing hydrogen attack. Etch: picral

diffusion rate of carbon at the low operating temperature of the cooler. However as shown earlier [1]), the transformation of a very small amount of dissolved carbon is sufficient to initiate the formation of fissures at the grain boundaries by means of the methane reaction. The attack penetrated to a depth of 13 mm under the longitudinal boring in the broken manifold; a penetration depth of up to 16 mm was measured in other manifolds. The depth of attack decreased from the gas inlet side to the outlet side. Figures 9 to 11 show hydrogen attack of varying depth and intensity on unetched polished specimens on

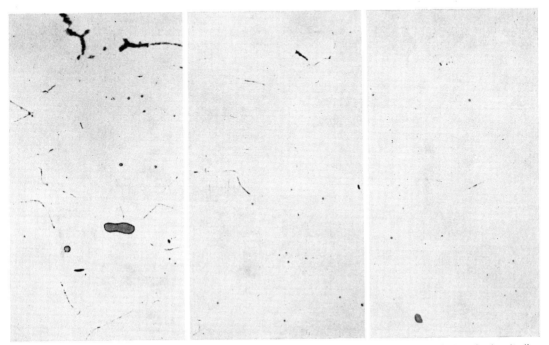

Fig. 9. 0.3 mm below the longitudinal boring

Fig. 10. 11 mm below the longitudinal boring

Fig. 11. 14 mm below the longitudinal boring

Figs. 9 to 11. Hydrogen attack at the inlet side of the upper manifold, unetched. 500 x which this effect can be verified most reliably.

Hydrogen attack always results in a decrease in tensile strength; in particular it causes an extensive loss of ductility. DVM notched impact specimens, which were sectioned parallel to the crossboring with the notch perpendicular to the boring from the attacked inner region, and similar samples taken from the unaffected outer region, yielded the following values:

Specimen location	Specific impact energy [kg m/cm²]	
inner region	0,9	1,1
outer region	6,1	7,3

An examination of the cooler-tubes did not reveal any attack. Here the internal temperature was evidently lower because of the lesser wall thickness.

This investigation has shown that the unalloyed steel used is not durable under the existing operating conditions. At least the cooler manifolds should therefore be replaced by manifolds made of hydrogen resistant steel [2].

---

[1] F. K. NAUMANN, F. SPIES, Prakt. Metallographie 8 (1971) 375/84

[2] Werkstoff-Handbuch Stahl u. Eisen, 4. Aufl., Blatt 0-85

Rivet-Hole Cracks in a Steam Boiler

Friedrich Karl Naumann and Ferdinand Spies

Max-Planck-Institut für Eisenforschung
Düsseldorf

Steels with nitrogen, when they are under mechanical stress, could be corroded with intercrystalline cracking by hot lyes and alkaline or slightly acid brine, especially if the steels have low carbon content [1]. The solution must have such composition that a potential builds up on iron, resulting in a partial passivity of certain parts of the microstructure, here the grain faces, and a high concentration of corrosion-current at the grain boundaries [2]. These conditions can be encountered in such harmless electrolytes like boiler feed water, when there is a concentration increase or differential aeration in narrow slits. Hot nitrate solutions are very corrosive. Therefore this solution is used to test steels for caustic embrittlement [3]. Internal stresses, especially in cold-worked zones are also taken into account.

The classic example of an intercrystalline stress corrosion is the well known rivet-hole cracking in steam boilers or brine-evaporators. After going over to welded construction and introduction of caustic-cracking resistant steels by A. Fry [4], rivet-hole cracks in steam boilers are not often encountered. In the following this type of failure is to be illustrated in an example of a water tube boiler with two headers and 15.5 at working pressure from the year 1913, which was examined in 1926 in the Kaiser-Wilhelm-Institut für Eisenforschung [5]. The boiler became leaky in the

lower part due to the formation of cracks in the rivet-hole edges. In the steam room such damages were not observed. The boiler plate of 20 mm thickness was a rimming steel with 0.05 % C, traces of Si, 0.38 % Mn, 0.027 % P, 0.035 % S, and 0.08 % Cu. The mean value of the yield point was 24 (24) kg/mm², the tensile strength 39 (38) kg/mm², the elongation at fracture, δ_{10}, 26 (24) %, the necking at fracture 71 (66) % and notch impact value 11,5 (9,4) kgm/cm² (the values in brackets are for the transverse direction). The direction of rolling was parallel to the rivet seam, as confirmed by a micro structural examination. Normalizing did not affect the good mechanical properties.

Figures 1 and 2 show the place of failure and the surrounding zone. The inside surface of the boiler was polished and etched with Fry-solution, so that the cold working could be seen. In the right hand side of Fig. 1 and its continuation in Fig. 2 parallel striations can be seen, which are due to the cold bending of the plate. On the left of these striations (in Fig. 1 on the right) a darkly etched band, which runs parallel to the rivet seam can be seen. This deformation is due to the caulking of the plate edge. The zones of slip are concentrated around the rivet holes, where the plate material is cold worked by the pressing of rivet heads. The cracks are formed here.

Figure 3 shows the micrograph of a section vertical to the parallel striations of Fig. 2 etched with Fry-solution. This etching proves that the deformation due to cold bending has reached the middle of the plate from the tensile as well as the compression zone. Figures 4 and 5 show similarly etched micrographs through the double row of rivet seam. The "stretcher lines" have their highest density and depth at the hole rims. The deforma-

[1] Ed. HOUDREMONT, H. BENNEK, H. WENTRUP, Erforschung und Bekämpfung der interkristallinen Korrosion des unlegierten Stahles, Stahl u. Eisen 60 (1940) 757/63 u. 791/801

[2] K. BOHNENKAMP, Über die Spannungsrißkorrosion weicher Stähle in konzentrierter Natronlauge. Arch. Eisenhüttenwes. 39 (1968) 361/68

[3] Stahl-Eisen-Prüfblatt 1860, Verlag Stahleisen mbH, Düsseldorf

[4] A. FRY, Das Verhalten der Kesselbaustoffe im Betrieb, Krupp'sche Monatshefte 7 (1926) 185/96

[5] A. POMP, P. BARDENHEUER, Schadensfälle an Dampfkesselelementen, Mitt. K.-Wilh.-Inst. für Eisenforsch. 11 (1929) 185/91

1st row of rivet holes   2nd row of rivet holes

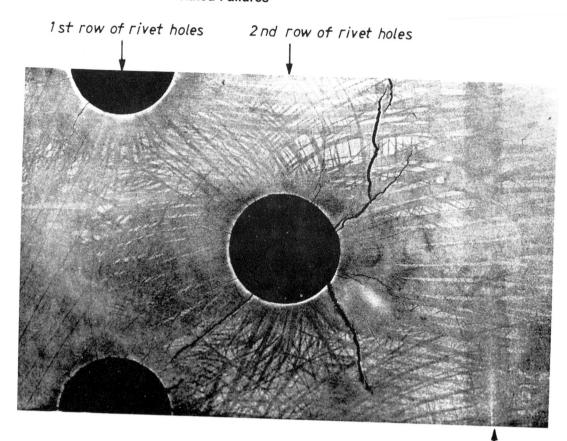

Fig. 1. Inner surface around rivet holes with cracks

caulked edge

Fig. 2. Continuation of Fig. 1 at the right

Figs. 1 to 3. Etched with Fry-solution. 1 x

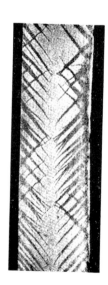

Fig. 3. Micrograph parallel to rivet seam in the zone of Fig. 2

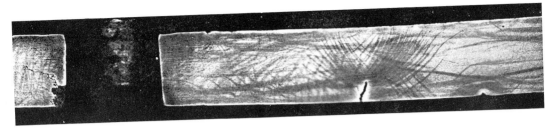

Fig. 4. Transverse micrograph through a rivet hole in the first row

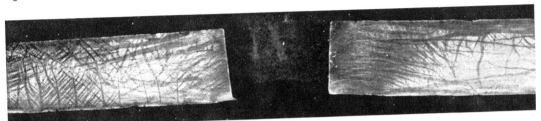

Fig. 5. Transverse micrograph through a rivet hole in the second row

Figs. 4 and 5. Etched with Fry-solution. 1 x

tions due to caulk seam reach deep into the material.

Even from the fact that cracks have only formed in the water room of the boiler one can conclude that they have resulted from corrosion, which was confirmed by the structure examination. Figures 6 and 7 prove that the cracks have taken an exactly intercrystalline path. As said before this is characteristic for caustic corrosion cracks.

At the end a couple of words as to what could be done on the material side against such intercrystalline stress corrosion. As first the manufacture and processing should be as such that no unnecessary mechanical or thermal residual stresses remain. Existing internal stresses should if possible be removed through annealing. Where it is neither possible nor advisable a lye-resistant steel should be used. The "Izett"-steel, developed by A. F r y in the Fried. Krupp works in Essen at the beginning of the 1920s, which was deoxidized (denitrated) with aluminium, has a high resistance against intercrystalline stress corrosion, which can even further be improved by increasing the Al-content.

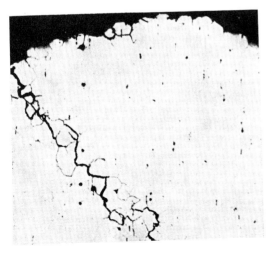

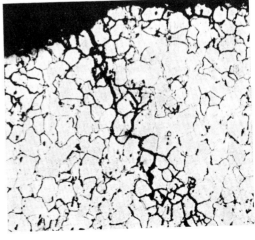

Fig. 6. Unetched

Fig. 7. Etched with nital

Figs. 6 and 7. Crack path. 100 x

Examination of Corroded Boiler Tubes

Friedrich Karl Naumann and Ferdinand Spies

Max-Planck-Institut für Eisenforschung
Düsseldorf

Two pipe sections with attacked areas in the circumferential welding joint were received for examination. The sections originated from a backwell tube situated in the combustion chamber of a 100-atm. boiler, which had been in service 4½ years. The temperature of the saturated steam was about 300° C. It is possible that the tube temperature reached 400° C. This investigation was to reveal whether the attacked areas can be attributed to corrosion phenomena or whether we are dealing with welding burns.

One of the sections, henceforth designated No. 1, was severed longitudinally and cut transversely in the plane of the seam. This specimen shows strong pit- or trench-like attack in the welding seam on the inner surface. A bluish-black corrosion product adhered to the pits which was in clear contrast with the red coating on the rest of the inner surface. The other pipe section (No. 2) also contained a circumferential welding joint. The cross-sectional cut was situated approx. 10 mm from the bead. After the section was cut longitudinally, the welding seam, in contrast to the first specimen, did not show clearly discernible attacked areas but only single smaller blisters whose nature could not be established externally. The welding material had in places penetrated to the interior and formed a burr on the inner surface of the tube.

For the purpose of metallographic examination of the first section, the available transverse section through the plane of the welding seam was prepared. In addition, three longitudinal sections through locations of different intensities of attack were made. Only one longitudinal section of the second pipe section was investigated.

Figures 1 and 5 show slightly magnified longitudinal and transverse sections through the region of deepest corrosion penetration

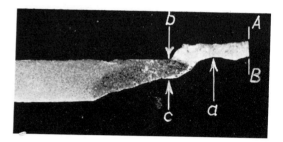

Fig. 1. Specimen 1, Longitudinal section, etch: picral. 2 x
(top: outside, bottom: inside, right: weld)

of the strongly-attacked pipe section. The welding seam is strongly reduced in bulk from the inside and covered with a thick crumbling layer of magnetic iron oxide (Fe_3O_4) — see Fig. 2. The oxide layer is much thicker than the external layer of scale. This fact alone led to the conclusion that we were not dealing with welding scale but with a corrosion product resulting from the operation of the boiler. The region next to the seam, which was overheated during welding, had also been attacked. The attack ends fairly precisely where the normal grain size of the pipe material begins again. The weld seam was very coarse-grained (Fig. 2), and also the transition zone was strongly overheated and in places the grain boundaries were oxidized due to burning (Fig. 3). In addition, it was decarburized from the inside, and interspersed with grain boundary cracks (Fig. 4). This form of attack is typical for the decarburization of steel by high-pressure hydrogen [1]. The intensity of attack decreased with distance from the weld seam in the longitudinal section, and similarly, with distance from the region of

[1] F. K. NAUMANN, F. SPIES, Prakt. Metallographie 8 (1971) 375/384

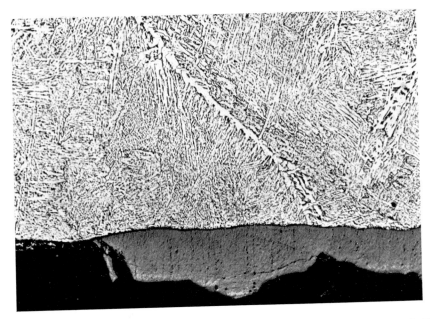

Fig. 2. Structure at inside boundary of weld seam (Location a in Fig. 1), etch: picral. 100 x

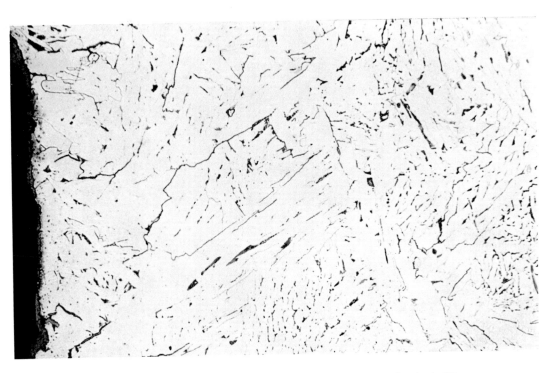

Fig. 3. Structure at outside boundary of transition zone (Location b in Fig. 1), etch: picral. 100 x

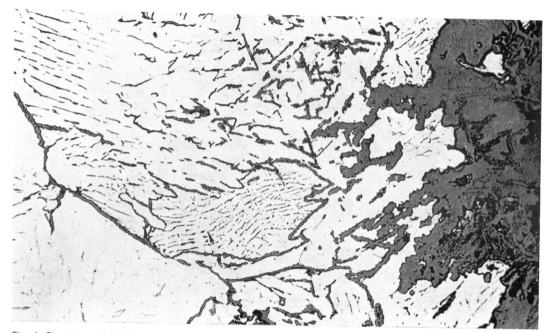

Fig. 4. Structure at inside boundary of transition zone (Location c in Fig. 1), etch: picral. 100 x

Fig. 5. Transverse section A – – B see Fig. 1, etch: picral. 2 x

pit formation in the transverse section (Figs. 6 to 8). Right next to the seam, the pipe material was completely decarburized and the crack-edges were oxidized (Fig. 4). In this region the attack encompassed almost the total wall thickness. Further away from the corrosion pit, the depth and intensity of decarburization as well as the number

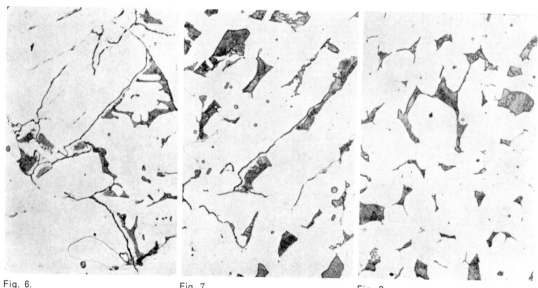

Fig. 6.   Fig. 7.   Fig. 8.

Figs. 6 to 8. Structure in the interior part of the pipe wall with increasing distance from the location of corrosion pitting. 200 x

and width of cracks decreased; likewise the grain size also becomes smaller. The attack finally only becomes discernible by a small widening of the grain boundaries and the fissured grain boundaries can only be distinguished from normal grain boundaries in the unetched condition or after a slight etch with picric acid in alcohol.

Figure 9 reproduces the longitudinal section through the less strongly attacked pipe section 2. The welding material which had penetrated to the inside of the pipe, was strongly interspersed with blisters and coarse roundish oxide inclusions. Nevertheless, attack by scaling was not detectable macroscopically neither in the weld seam nor right next to it. Only at some distance from the seam has a small corrosion pit been cut (location a in Fig. 9). Here the surface was covered with a thick layer of scale, and the structure under this layer clearly showed the signs of hydrogen attack (Fig. 10), which at this location had already encompassed approximately 2/3 of the wall thickness. Towards the weld seam beyond the corrosion pitting, the intensity and depth of attack decreased suddenly. On the other side of the weld seam no attack was visible even in close proximity of the weld. It is noteworthy that also the grain size was smaller on this side than on the other. On the whole, this weld seam was appreciably finer-grained in its interior part than the one examined first; this is also true for the bordering regions of the pipe material. This weld was probably placed in two passes and the grain size of the inner layer was changed during the deposition of the second bead.

According to this finding it is certain that the defects in the pipe sections are the result of scaling during the operation of the steam boiler and cannot be attributed to burning during the welding process. The following series of events appears to have occurred: At first, magnetic iron oxide and hydrogen were formed in accordance with the equilibrium reaction according to Chaudron:

$$3\,Fe + 4\,H_2O = Fe_3O_4 + 8\,H$$

The reason why the weld joints and the region surrounding them were preferentially attacked by the reaction products may be one of the following: The leafy scale formed during the welding operation prevented the development of a firmly adherent protective coating. Another possibility is that such a coating did develop and may subsequently have been destroyed by the turbulence

Fig. 9. Overall view. 2 x

Fig. 10. Structure in region of corrosion pit a in Fig. 9. 200 x

Figs. 9 and 10. Specimen 2, longitudinal section, etch: picral

formed at the protruding ridges of the weld joint. The precisely defined pit-like shape of the attacked areas indicates localized damage to the protective layer. The penetration of oxygen at the grain boundaries may have been facilitated by the manifestations of burning in the adjacent overheated region. The hydrogen resulting from the decomposition of steam then diffused into the steel, decarburizing it with the formation of methane and causing the crack formation at the grain boundaries. Fresh surfaces for attack by scaling reaction were created in the interior of the structure as a result of the loosening of the structure (see Fig. 4). A structure loosened in this manner should

also have a diminished erosion resistance. The original scaling process may in this way have been accelerated by hydrogen attack and mechanical removal of material.

This investigation has shown the following: It is more important than usual to avoid unnecessary overheating or unduly long procedures during the welding of materials for high-pressure steam boiler operations. Care has to be taken that no ridges or burrs form which might interfere with the development of laminar flow and a dense layer of scale. If possible, the weld seams and adjoining regions should be normalized.

Leaky Heating Coils of an Austenitic Chromium-Nickel-Molybdenum Steel

Friedrich Karl Naumann and Ferdinand Spies
Max-Planck-Institut für Eisenforschung
Düsseldorf

A solution containing 50 to 70 % calcium chloride, with a pH value of 7.5 to 8.5, was concentrated by evaporation in a brick-lined vessel by passing steam at a pressure of 15 atmospheres through a system of heating coils made of stainless austenitic steel X 10 CrNiMoTi 18 12 (material No. 1.4573). The final temperature of the solution was about 170° C. After five months one of the coils, which consisted of tubes having an outside diameter of 68 mm and a wall thickness of 3.4 mm, developed a leak.

No removal of material due to corrosion could be detected either on the inside or the outside of the tube. However, indications of tightly closed cracks could be seen on the outer surface, which had been in contact with the chloride solution. A test for cracks by the magnetic powder technique was not possible with the austenitic material. By means of the colour penetration process, however, numerous, multiple branched cracks could be revealed (Fig. 1).

In a longitudinal section, Fig. 2, it was seen that the cracks had started from the outside surface of the tube. They have propagated mainly across the grains. After electrolytic grain boundary etching in aqueous nitric acid, Fig. 3 a, this can be seen particularly clearly. Twin boundaries also fail to deflect the cracks from their original direction, which is determined solely by the stress (Fig. 3 b).

These observations indicate that this is a typical case of transcrystalline stress corrosion. Stress corrosion cracks are formed in metallic materials susceptible to stress corrosion, under the influence of specific corrosive media, only when they are simultaneously subjected to tensile stresses. The cracks can propagate across the grains or along grain boundaries. All austenitic chromium-nickel alloys are susceptible to the type of transcrystalline stress corrosion observed in the present case. The susceptibility decreases with increasing stability of the

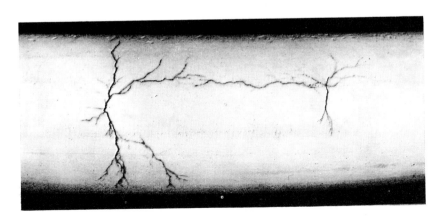

Fig. 1. Surface of the heating coil. Cracks shown up by the colour penetration process. 0.5 x

Fig. 2. Cross-section through the cracked region, unetched. 10 x

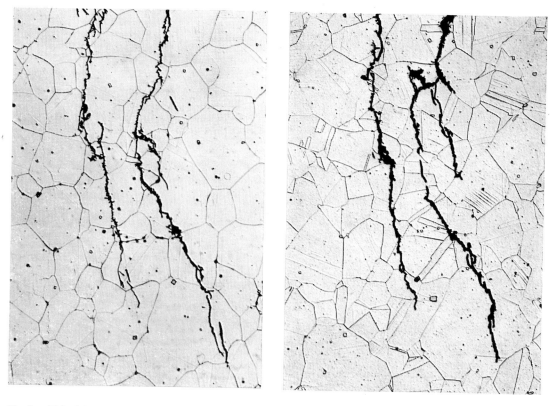

Fig. 3 a. Etched in aqueous nitric acid 1:1 at 2 V for 3 min.

Fig. 3 b. Etched in V 2 A pickling solution (50 ° C)

Figs. 3 a and 3 b. Crack propagation, cross-section (Same area after different etching treatments). 200 x

austenite, and therefore with increasing nickel content. Austenitic chromium-manganese steels and ferritic chromium steels are considerably more resistant. As a rule, however, these steels are unsuitable as replacement because of their greater sensitivity to pitting. Specifically effective corrosive media are chlorides such as calcium and magnesium chloride, but also highly concentrated alkali solutions such as potassium or sodium hydroxide. The process takes place very slowly at room temperature, but is strongly accelerated by raising the temperature. Stresses are effective even if they lie far below the yield point of the material. They can be external stresses, e. g. working stresses, or internal stresses, e. g. deformation or welding stresses. The mechanism of stress corrosion is not yet fully understood.

[1] H. J. ROCHA, Zur Spannungskorrosion austenitischer Stähle. Techn. Mitt. Krupp, Forschungsberichte 5 (1942) 1/14

[2] H. KAESCHE, Die Korrosion der Metalle, Springer-Verlag, Berlin (1966)

Fractures of Electro-Galvanized Cylinderhead Screws

Friedrich Karl Naumann and Ferdinand Spies
Max-Planck-Institut für Eisenforschung
Düsseldorf

Eight 7/16" cylinderhead screws which had cracked after a short running time of the motors were submitted for examination. They were made of a steel containing 0.45 % C and 1 % Cr, had rolled threads, were heat treated to approx. 110 kg/mm² tensile strength, and were electrolytically galvanized. All were fractured at the root of the thread. The surfaces of fracture were fine-grained and not spread by rubbing. They are judged to be forced ruptures or perhaps fatigue fractures which had progressed rapidly.

Figure 1 shows a longitudinal polished section through the failure location of a screw. Apart from the crack which led to fracture (the origin of the fracture is marked by an arrow in Fig. 1), additional starting cracks in several threads, and in part even a number of incipient cracks close together in the same thread, were noted (Fig. 2). The cracks proceeded in the jagged and branched manner typical of tension cracks (Figs. 3 and 4).

Pattern and course of the cracks do not exclude the possibility that they may be due to fatigue failure, but in view of the fact that the screws were electrolytically galvanized, they more strongly indicate that we may be dealing with so-called "delayed fracture" in the failure of the screws. Experience has shown that this type of fracture is seen on production parts made of high-strength

Bruchausgang/crack origin

Fig. 1. Longitudinal polished section through the thread of a ruptured screw, unetched. 7 x

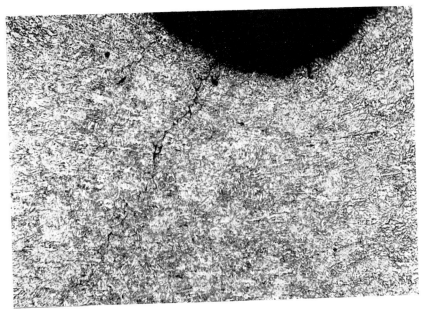

Fig. 2. Fine incipient cracks originating from the root of the thread. Longitudinal section, etch: nital. 200 x

Fig. 3. Crack origin in the root of thread

Fig. 4. Course of crack

Figs. 3 and 4. Crack in Fig. 1 at higher magnification, longitudinal section, etch: nital. 200 x

steels, which have absorbed hydrogen during pickling or during a galvanic surface treatment. Such parts will rupture below the elastic limit during continuous stressing. This will often occur only after the expiration of a certain time period, and preferably at locations of stress concentrations such as changes in cross-section or threads [1]). As a rule, the hydrogen cannot be verified analytically since most of it escapes again after prolonged storage at room temperature or short heating at 100 to 200° C.

---

[1]) Vgl. H. KRAINER, W. REICH, Werkstoff-Handbuch Stahl u. Eisen, 4. Aufl., Blatt G 12

Damage to Tool Joints in Hydrogen Sulphide–Carrying Natural Gas Drilling Operation

Friedrich Karl Naumann and Ferdinand Spies

Max-Planck-Institut für Eisenforschung
Düsseldorf

During drilling for natural gas in the Ems region in the year 1956, considerable amounts of longitudinal cracks and transverse fractures occurred in the connecting pieces of the bore rods. The connectors, whose shape can be seen in Fig. 1, were screwed onto the rods by means of a fine thread and tightly joined with it by shrinkage at 530° C. The connectors were made of chromium-molybdenum steel SAE 4140 with approx. 0.4 % C, 1 % Cr and 0.2 % Mo, and they were hardened and tempered to 90 to 105 kp/mm² tensile strength. The material for the rod pipes was steel with 0.4 % C and 1 % Mn of 68 kp/mm² strength.

Visual inspection showed that 87 out of 172 dismantled connecting parts were destroyed, that is, over one-half. The amount of coupling- and pin cracks were about even. The cracks appeared predominantly in the heat shrunk fine thread of the pins and couplings (Fig. 2), but in some cases also in the coarse thread part of the couplings (Fig. 3). This negated the assumption that shrinkage stresses were the cause of the damage,

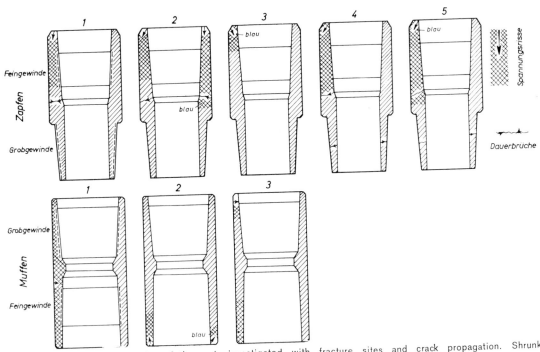

Fig. 1. Schematic representation of the parts investigated with fracture sites and crack propagation. Shrunk connectors 4½″

→ fatigue fracture

Fig. 2. Pin (5) with longitudinal crack and fatigue fracture in fine threaded part and transverse fracture in coarse thread. 0.3 x

Fig. 3. Coupling with longitudinal cracks in fine and coarse threaded part. 0.3 x

even though the fractures were tarnished blue.

As far as could be ascertained by external observation, the cracks originated in the end faces, i. e. the points of highest shrinkage or screw stresses. The longitudinal cracks at the pins were usually upon attaining a certain length transformed immediately or gradually into a circumferential transverse crack (Fig. 2). Some pins also showed transverse fractures in the coarse thread (Fig. 2).

No fractures or cracks could be observed at the bore rods. It was subsequently established that steels of lower strength are con-siderably less sensitive to fractures of the kind considered here.

It is noteworthy that the damage appeared only in a certain well. Its cause, therefore, was bound to lie in the specific conditions of this well. Subsequently it was found that the natural gas drilled in these locations had an exceptionally high hydrogen sulphide content that even after passing through the scrubber still amounted to 5 to 6 %, and therefore must have been considerably higher originally.

Five pins and three couplings with especially significant aspects were selected for

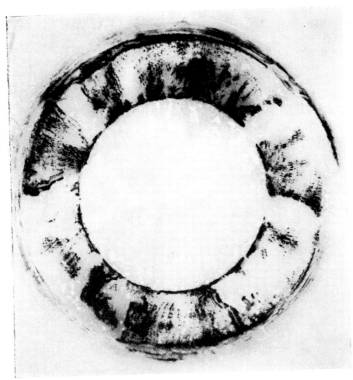

Fig. 4. Sulphur print according to Baumann from fracture plane of a transverse fracture (reproduction). approx. 0.5 x

Fig. 5. Pin 5, not cleaned at left, pickled at right. 0.5 x

Fig. 6. Pin 3, not cleaned at left, pickled at right, 1 x

Figs. 5 and 6. Pin cracks in the fine threaded part, opened up (crack initiation designated by arrows)

investigation. Chemical composition and strength properties corresponded to specifications. To seek out the cause of fracture it was necessary, as always, to determine the point of origin of the cracks. For this purpose all cracks were opened up. The fracture planes were in part tarnished blue. By means of a sulphur print according to B a u m a n n (Fig. 4) it was later established that this was not an oxidic tarnish, but a coating of iron sulphide. The points of origin and the fracture propagation are designated by arrows in Fig. 1. In the pins and coupling 2, the cracks originated in the face plane of the fine threaded part (Fig. 5). In pin 3 a deviation may be noticed since the crack runs not along the inner edge which theoretically is exposed to the highest stress, but

Fig. 7. Coupling 3, not cleaned at left, pickled at right.
1 x

Fig. 8. Coupling 1, pickled. approx. 0.6 x

Figs. 7 and 8. Coupling cracks, opened up (crack initiation designated by arrows)

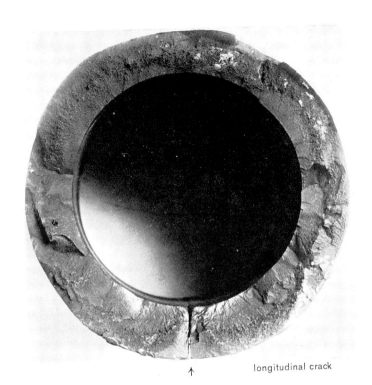

↑
longitudinal crack

Fig. 9. Fatigue fracture in fine threaded part of pin 1. In area of longitudinal cracks erosion through exuded fluid. 0.5 x

along the outer rim (Fig. 6). However, an entirely different fracture propagation appears in the couplings 1 and 3. Here the cracks clearly originated at the exterior. In coupling 3 this was just opposite the large shrink fit in the fine threaded part and opposite the seal plane and the first thread turn in the coarse threaded part (Fig. 7). In coupling 1 this was still farther away from the face plane opposite the small shrink fit (Fig. 8). The unusual fracture origin from points of small mechanical stress pointed to other origins of the stresses that may also have played a role in the other connectors.

The transverse cracks in the fine threaded part, however, proved to be fatigue fractures that ran from the small shrink fit to the seal plane and also in part and probably secondary-originated at the opposite exterior plane (Fig. 9, compare with arrows in Fig. 1). The fatigue cracks are probably caused secondarily by alternating bending stresses that occurred after the tearing open of the pins by losening of the slip joint. The transverse cracks in the coarse threaded part of the pins are internally initiated stress cracks.

In coupling 1 with continuous longitudinal crack and external fracture origin (compare Fig. 8), a number of deep seated longitudinally directed impression marks were noticeable, that were caused by the pliers used for tightening of the screws. In order to find out whether the unusual fracture origin at the external plane had any connection with the plier marks, several sections were made across these points. The steel that showed a fine-grained and fine-needled heat treated structure was deeply cold deformed under the plier marks (Fig. 10) and fractured in the direction of shear stress. Thus the fracture formation was helped along by deformation stresses.

A different kind of crack occurred at several microsections without any connection to plier marks. These are short stress cracks and are similar to flakes in whose formation hydrogen plays a part[1]. They originate internally like the former, and all end 1/2 to 1 mm under the surface (Fig. 11). The fracture of the coupling may have started from such cracks which continue in a radial direction, in part over one-third to one-half of the wall thickness.

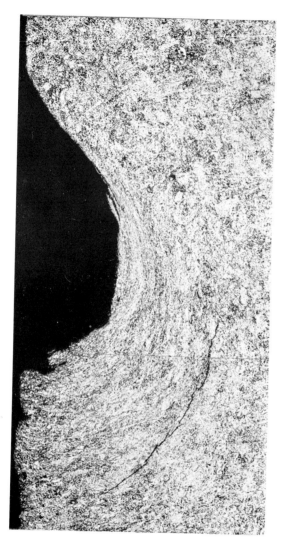

Fig. 10. Structure under a plier mark in coupling, transverse section, etch: Nital. 65 x

Fig. 11. Interior stress crack in coupling 1, transverse section, etch: Nital. 60 x

Initiation and propagation of these cracks point to the fact that structural stresses play a role in their appearance, and in part they have the principal role in the sum of the stresses. From the investigation — formation of iron sulphide on the fracture planes and flake-like stress cracks in the steel — it could be concluded that the hydrogen sulphide content of the gas was the true cause of the damage. It could be further concluded that the hydrogen that was liberated by reaction with the iron causing the formation of iron sulphide after penetration of the steel, had an explosive effect during molecular separation under high pressure. This in turn caused the crack formation in conjunction with the external and residual stresses. This assumption was confirmed by stress corrosion tests with bend specimens (Jominy specimens) [2] that used the same corrosive solution that had been used for the flushing during drilling. The solution was used once in the state in which it came out of the well and again after boiling to remove the hydrogen sulphide. Under high stresses those samples of the steel used for the connector that were in the original solution fractured already on the first day, while those in the boiled solution still stood up after 128 days. The fracture-promoting effect of a cold deformation was confirmed as well. Specimens made of the softer rod steel did not fracture even after 86 days in more effective saturated hydrogen sulphide water.

Subsequently the mechanism of the process was thoroughly investigated [3].

Generally valid recommendations for the avoidance of such damage could not be given with respect to the material. Keeping the strength low is not possible in view of the mechanical stress. A chromium addition of a level for customary alloyed structural steels is useless and even a higher alloy content provides no absolute protection. Metallic or non-metallic coatings are either not wear-resistant or they are brittle and porous. It appears to be more successful to lower the attack potential of the hydrogen sulphide solution by raising its basicity or by the addition of inhibitors.

[1] H. BENNEK, H. SCHENK, H. MÜLLER, Stahl u. Eisen 55 (1935) 321/331

[2] F. K. NAUMANN, Erdöl-Zeitschr. 73 (1957) 4/14

[3] F. K. NAUMANN, W. CARIUS, Arch. Eisenhüttenwes. 30 (1959) 233/238; 383/392; 361/370

Cracked T-Piece from a Copper Hot Water System

Egon Kauczor
Staatliches Materialprüfungsamt an der Fachhochschule
Hamburg

The site of the damage in the T-piece from a copper hot water system is shown magnified in the general view in Fig. 1. A specimen was taken from the vicinity of the crack for metallographic investigation.

Microscopic examination of the polished section revealed, in addition to the main crack visible in Fig. 1, further branching transcrystalline cracks running from the outer surface of the pipe into the pipe wall. The micrograph in Fig. 2 of part of the polished specimen shows as well as cracks, the homogeneous copper matrix, a more darkly etched edge zone and, top left, streamers from the soldered joint.

This appearance indicates disintegration by stress corrosion cracking. Although copper is not susceptible in the pure state, it has a tendency to stress corrosion cracking under tensile stress in the presence of very small quantities of other elements in a damp ammoniacal atmosphere. A susceptibility was detected at only 0.004 % phosphorus. The greatest sensitivity lies between phosphorus contents of 0.028 and 0.46 % [1] [2].

The chemical analysis of the pipe material yielded a phosphorus content of 0.028 % within the critical range. The material is, however, not defective but a phosphorus-deoxidized copper type. According to DIN

Fig. 1. General view of the site of the damage with a crack at ↓. 2 x

Fig. 2. Polished specimen from the site of the fracture etched with ammonium persulphate showing branched transcrystalline cracks lying to the right of the main crack visible in Fig. 1. 200 x

1708, phosphorus-deoxidized copper should have a residual phosphorus content of 0.015 to 0.05 %. This residual phosphorus combines with oxygen to form phosphorus pentoxide (darkly etched oxidation zone in Fig. 2) when the copper is processed at high temperatures, e. g. hard soldered, thereby preventing the formation of cuprous oxide and hence the readily occurring hydrogen embrittlement.

[1] K. DIES, Kupfer und Kupferlegierungen in der Technik, Springer-Verlag, Berlin, Heidelberg, New York (1967)

[2] D. H. THOMPSON, A. W. TRACY, Trans. AIME 185 (1949) 100/109

Pressure Vessel from a High-Pressure Vibratory Autoclave Burst by Explosion

Egon Kauczor

Staatliches Materialprüfungsamt an der Fachhochschule
Hamburg

The damaged forged pressure vessel made from high temperature austenitic steel X8CrNiMoVNb 16 13 K (material no. 1.4988) is shown in Fig. 1 as it was on arrival. Closer examination of the widest part of the burst revealed fine cracks on the internal wall running in a longitudinal direction. When the internal wall was cleaned, numerous even finer cracks were exposed as shown on the section in Fig. 3. On the fracture surfaces in this region an irregularly formed zone was visible in the direction of the internal wall and a fibrous oriented fracture zone towards the external wall (Fig. 4).

Further very fine branched cracks were revealed on microsections taken for metallographic examination from the cracked region shown in Fig. 3. The micrograph in Fig. 5 shows such a crack in an unetched microsection. On the more highly magnified micrograph of the etched section (Fig. 6) the transgranular character of the cracks is clearly visible.

This picture is typical of stress corrosion cracking in austenitic steels.

Stress corrosion cracking can occur when the following three factors are present:

1) A material susceptible to stress corrosion cracking

2) A specific corrosive agent capable of causing stress corrosion cracking in this material

3) tensile stresses

The susceptibility of austenitic steels to transgranular stress corrosion cracking in the presence of chlorides is well known. In

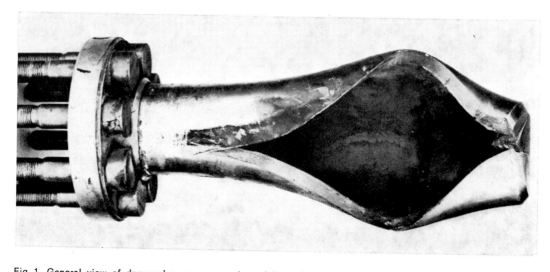

Fig. 1. General view of damaged pressure vessel as delivered. ¼ x

Fig. 2. Cracks in the internal wall at the widest part of the burst. 1 x

Fig. 3. Part of the cracked zone after cleaning. 10 x

Fig. 4. Enlarged section of fracture edge at the widest part of the burst. 5 x

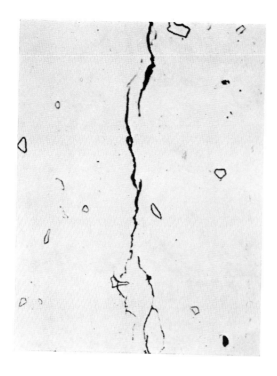

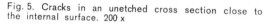

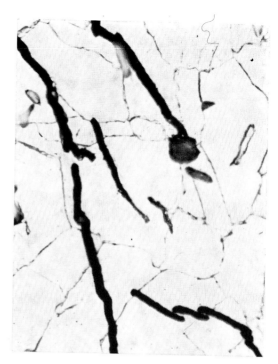

Fig. 5. Cracks in an unetched cross section close to the internal surface. 200 x

Fig. 6. Highly magnified micrograph of section etched with stainless steel pickle. 1000 x

the experiment, which, according to the accident report, led to the explosion, vanadium trichloride was used and tensile stresses were of necessity set up by the internal pressure.

It is however scarcely possible that this experiment alone caused the entire damage. It is far more likely that in the course of the years in numerous similar experiments the vessel wall was weakened by increasingly numerous stress corrosion cracks so that gradually it was no longer capable of withstanding the normal experimental conditions for which the pressure vessel was constructed.

It can also be seen on the section from the fracture edge in Fig. 4 that more than half the wall was broken up by stress corrosion

cracking even before fracture occurred. The fibrous fracture surface (right) follows the impurities lying in the forging direction and the similarly oriented carbides of the stabilizing element. The larger internal fracture surface on the other hand has followed the previously existing stress corrosion cracks.

Stress corrosion cracking does not occur if one of the basic requirements is lacking. Since in the present case the agent and tensile stresses are inevitably present the only possibility is to forge the whole pressure vessel from a material which is not susceptible to stress corrosion cracking or to use interchangeable linings of such a material. A nickel alloy could for example be considered.

Broken Exhaust Valve

Rudolf Kallenbach and Rudolf Fichter

Eidgenössische Materialprüfungs- und Versuchsanstalt
Dübendorf

The object under examination was a broken exhaust valve with disk diameter 30 mm and stem diameter about 8 mm from the cylinder of a motor car. The site of the fracture is directly where the valve cone joins the cylindrical stem. Both the cone and the stem are heavily scaled in the vicinity of the fracture; in some parts the scale has flaked off. Furthermore the rim of the disk is badly damaged by secondary mechanical action.

Axial sections taken transversely through the fracture surface reveal that the core of the valve has a very fine austenitic microstructure with precipitations of numerous granular and very fine, mostly rounded carbides and fine segregation bands. In the head the fibres are in some cases bent over. Above all the edge zone of the cone and the upper part of the stem indicate a thermal effect as a result of which the fine carbide precipitates are formed somewhat differently. A hard alloy facing has been welded on to the valve seat.

A fringelike oxide-rich diffusion zone about 0.02 mm thick has formed in the edge zone of the cone and adjoining stem and in the edge zone localised grain boundary oxidation has occurred. Numerous fine, quite straight transverse cracks with a fine oxide border and enclosed oxide particles have formed in the stem (Fig. 1). There is a layer

Fig. 1. Longitudinal section through the fracture, unetched. Oxide rich seam in the edge zone; cleft like crack with enclosed oxide. Fracture site to the left. 200 x

Fig. 2. Microprobe: nickel distribution. 625 x

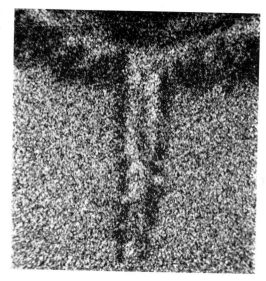

Fig. 3. Microprobe: manganese distribution. 625 x

of scale on the surface that reaches a thickness of 0.15 mm at the transition between the stem and the cone. This scale layer is cracked in places and in some cases the cracks in the scale are continued as fine cracks in the metal.

The corrosion phenomena (scaling) in particular were studied in the vicinity of a fine cleft shaped crack in the stem with the microprobe. It can be seen from the electron picture and the oxygen and iron distribution pictures that spots of oxide precipitate are present in the edge zone and crack walls. The chromium distribution picture indicates enrichment in the edge zone and at the crack and deficiency in the adjacent (to some extent richer in nickel) zones. The edge zone and to a lesser extent the crack walls are enriched in nickel (Fig. 2). The nickel depleted regions are preferentially enriched in manganese (Fig. 3) which has a similar distribution to chromium and oxygen. Sulphur is distributed similarly to nickel in the edge zone and similarly to manganese along the cracks (Fig. 4) i. e. manganese sulphide has obviously formed alongside chromium oxide on the crack walls whereas the sulphur in the edge zone is bound predominantly to the nickel (nickel sulphide).

According to the results of the investigations, the exhaust valve is heavily scaled. The

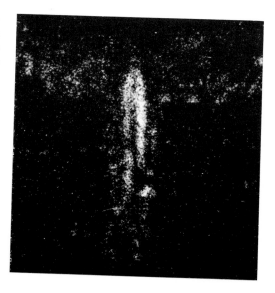

Fig. 4. Microprobe: sulphur distribution. 625 x

scale layer on the stem exhibits stress cracks by means of which the corrosion has penetrated more deeply leading to cracks in the metal. As a result of diffusion particularly of the alloying elements iron, chromium, nickel and manganese, various fringe-like enrichment zones have established themselves along the edges of the cracks in conjunction with diffusion inwards of oxygen and sulphur. The advance of locally initiated corrosion has in the course of time been promoted by external mechanical influences particularly vibrations and thermal stresses. The cracks have a characteristic, straight cleft like shape. Fracture is therefore a consequence of fatigue corrosion cracking, itself strongly promoted by the presence of sulphur compounds. The origin of these corrosive sulphur compounds could not be explained in the scope of the investigation.

Index